Essentials of Organic Chemistry

Essentials of Organic Chemistry

– Author –
Sanju Tomari

Wordpen Academics
Publishing Science That Matter

ISBN: 978-93-90438-38-9 (Hardback)

Published by : **WORDPEN ACADEMICS**
4736/23, Ansari Road, Darya Ganj,
New Delhi – 110 002
Phone: +91-011-35004336
E-mail: info@wordpenacademics.com

Printed at **:** **Replika Press Pvt. Ltd.**

Exclusive Distributor: ***Bio-Green Books, New Delhi***

Preface

Organic chemistry is that branch of chemistry that deals with the structure, properties, and reactions of compounds that contain carbon. It is a highly creative science. Chemists in general and organic chemists in particular can create new molecules never before proposed which, if carefully designed, may have important properties for the betterment of the human experience. In terms of Ph.D. population, organic chemistry is the largest chemistry discipline, in both total numbers, annual Ph.D. graduates, and in annual production.

Beyond our bodies' DNA, peptides, proteins, and enzymes, organic compounds are all around us. They are central to the economic growth of the U.S., in industries such as the rubber, plastics, fuel, pharmaceutical, cosmetics, detergent, coatings, dyestuffs, and agrichemicals industries. The very foundations of biochemistry, biotechnology, and medicine are built on organic compounds and their role in life processes. Most all of the modern, high tech materials are composed, at least in part, of organic compounds. Clearly, organic chemistry is critically important to our high standard of living. Organic chemists at all degree levels are found in all those industries that depend on R&D, working on projects from fundamental discovery to highly applied product development. The foundation of the pharmaceutical industry is the large pool of highly skilled organic chemists. For example, nature may provide a molecule such as a complex antibiotic, an antitumor agent, or a replacement for a hormone such as insulin; organic chemists determine the structure of this newly discovered molecule and then modify it to enhance the desired activity and specificity of action, while decreasing undesired side effects. Indeed, organic chemists have produced a wonderful myriad of highly successful products to fight human diseases. There is tremendous excitement and challenge in synthesizing a molecule never before made synthetically or found in nature. Tailoring the properties of that molecule via chemical synthesis to produce beneficial effects to meet the needs of the present and future human existence is both challenging and rewarding. The concept of functional groups is central in organic chemistry, both as a means to classify structures and for predicting properties. A functional group is a molecular module, and the reactivity of that functional group is assumed, within limits, to be the same in a variety of molecules. Functional groups can have decisive influence on the chemical and physical properties of organic compounds. Molecules are classified on the basis of their functional groups. Alcohols, for example, all have the subunit C-O-H. All alcohols tend to be somewhat hydrophilic, usually form esters, and usually can be converted to the corresponding halides. Most functional groups feature heteroatoms (atoms other than C and H). Organic compounds are classified according to functional groups, alcohols, carboxylic acids, amines, etc. Biomolecular chemistry is a major category within organic chemistry which is frequently studied by biochemists. Many complex multi-functional group molecules are important in living organisms. Some are long-chain biopolymers, and these include peptides, DNA, RNA and the polysaccharides such as starches in animals and celluloses in plants. The other main classes are amino acids (monomer building blocks of peptides and proteins), carbohydrates (which includes the polysaccharides), the nucleic acids (which include DNA and RNA as polymers), and the lipids. In addition, animal biochemistry contains many small molecule intermediates which assist in energy production through the Krebs cycle, and produces isoprene, the most common hydrocarbon in animals. Isoprenes in animals form the important steroid structural (cholesterol) and steroid hormone compounds; and in plants form terpenes, terpenoids, some alkaloids, and a class of hydrocarbons called biopolymer polyisoprenoids present in the latex of various species of plants, which is the basis for making rubber.

This book provides the chemical community with authoritative and critical assessments of the many aspects of organic chemistry. It is written for advanced undergraduate or beginning graduate student in chemistry with a sufficient foundation to comprehend and use the research literature in organic chemistry.

—Dr. Sanju Tomari

Contents

Chapter 1

Classification of Organic Compounds

Functional Groups

The concept of functional groups is central in organic chemistry, both as a means to classify structures and for predicting properties. A functional group is a molecular module, and the reactivity of that functional group is assumed, within limits, to be the same in a variety of molecules. Functional groups can have decisive influence on the chemical and physical properties of organic compounds. Molecules are classified on the basis of their functional groups. Alcohols, for example, all have the subunit C-O-H. All alcohols tend to be somewhat hydrophilic, usually form esters, and usually can be converted to the corresponding halides. Most functional groups feature heteroatoms (atoms other than C and H). The word moiety is often used synonymously to "functional group," but, according to the IUPAC definition, a moiety is a part of a molecule that may include either whole functional groups or parts of functional groups as substructures. For example, an ester (RCOOR') has an ester functional group (COOR) and is composed of an alkoxy moiety (-OR') and an acyl moiety (RCO-), or, equivalently, it may be divided into carboxylate (RCOO-) and alkyl (-R') moieties. Each moiety may contain additional functional groups—for example, methyl para-hydroxybenzoate contains a phenol functional group within the acyl moiety. Combining the names of functional groups with the names of the parent alkanes generates a powerful systematic nomenclature for naming organic compounds.

The atoms of functional groups are linked to each other and to the rest of the molecule by covalent bonds. When the group of covalently bound atoms bears a net charge, the group is referred to more properly as a polyatomic ion or a complex ion. Any subgroup of atoms of a compound also may be called a radical, and if a covalent bond is broken homolytically, the resulting fragment radicals are referred as free radicals.

The first carbon atom after the carbon that attaches to the functional group is called the alpha carbon; the second, beta carbon, the third, gamma carbon, etc. If there is another functional group at a carbon, it may be named with the Greek letter, e.g., the gamma-amine in gamma-aminobutanoic acid is on the third carbon of the carbon chain attached to the carboxylic acid group. Organic reactions are facilitated and controlled by the functional groups of the reactants. In general, alkyls are unreactive and difficult to get to react selectively at the desired positions, with few exceptions. In contrast, unsaturated carbon functional groups, and carbon-oxygen and carbon-nitrogen functional groups have a more diverse array of reactions that are also selective. It may be necessary to create a functional group in the molecule to make it react. For example, to synthesize iso-octane (the 8-carbon ideal gasoline) from the unfunctionalized alkane isobutane (a 4-carbon gas), isobutane is first dehydrogenated into isobutene. This contains the alkene functional group and can now dimerize with another isobutene to give iso-octene, which is then catalytically hydrogenated to iso-octane using pressured hydrogen gas.

Crystallography

The International Union of Crystallography in its Crystallographic Information File dictionary defines "moiety" to represent discrete non-bonded components. Thus, Na_2SO_4 would contain 3 moieties (2 Na^+ and one SO_4^{2-}). The dictionary defines "chemical formula moiety": "Formula with each discrete bonded residue or ion shown as a separate moiety".

Functionalization

Functionalization is the addition of functional groups onto the surface of a material by chemical synthesis methods. The functional group added can be subjected to ordinary synthesis methods to attach virtually any kind of organic compound onto the surface. Functionalization is employed for surface modification of industrial materials in order to achieve desired surface properties such as water repellent coatings for automobile windshields and non-biofouling, hydrophilic coatings for contact lenses. In addition, functional groups are used to covalently link functional molecules to the surface of chemical and biochemical devices such as microarrays and microelectromechanical systems.

Catalysts can be attached to a material that has been functionalized. For example, silica is functionalized with an alkyl silicone, wherein the alkyl contains an amine functional group. A ligand such as an EDTA fragment is synthesized onto the amine, and a metal cation is complexed into the EDTA fragment. The EDTA is *not* adsorbed onto the surface, but connected by a permanent chemical bond.

Functional groups are also used to covalently link molecules such as fluorescent dyes, nanoparticles, proteins, DNA, and other compounds of interest for a variety of applications such as sensing and basic chemical research.

Table of Common Functional Groups

The following is a list of common functional groups. In the formulas, the symbols R and R' usually denote an attached hydrogen, or a hydrocarbon side chain of any length, but may sometimes refer to any group of atoms.

Aliphatic Compounds

The aliphatic hydrocarbons are subdivided into three groups of homologous series according to their state of saturation:

- Paraffins, which are alkanes without any double or triple bonds,
- Olefins or alkenes which contain one or more double bonds, i.e. di-olefins (dienes) or poly-olefins.
- Alkynes, which have one or more triple bonds.

The rest of the group is classed according to the functional groups present. Such compounds can be "straight-chain", branched-chain or cyclic. The degree of branching affects characteristics, such as the octane number or cetane number in petroleum chemistry.

Both saturated (alicyclic) compounds and unsaturated compounds exist as cyclic derivatives. The most stable rings contain five or six carbon atoms, but large rings (macrocycles) and smaller rings are common. The smallest cycloalkane family is the three-membered cyclopropane ($(CH_2)_3$).

Saturated cyclic compounds contain single bonds only, whereas aromatic rings have an alternating (or conjugated) double bond. Cycloalkanes do not contain multiple bonds, whereas the cycloalkenes and the cycloalkynes do.

In organic chemistry, aliphatic compounds are acyclic or cyclic, non-aromatic carbon compounds. Thus, aliphatic compounds are opposite to aromatic compounds. In aliphatic compounds, carbon atoms can be joined together in straight chains, branched chains, or non-aromatic rings (in which case they are called alicyclic). Aliphatic compounds can be saturated, joined by single bonds (alkanes), or unsaturated, with double bonds (alkenes) or triple bonds (alkynes). Besides hydrogen, other elements can be bound to the carbon chain, the most common being oxygen, nitrogen, sulfur, and chlorine.

The simplest aliphatic compound is methane (CH_4). Aliphatics include alkanes (*e.g.* paraffin hydrocarbons), alkenes (*e.g.* ethylene) and alkynes (*e.g.* acetylene). Fatty acids consist of an unbranched aliphatic tail attached to a carboxyl group.

Properties

Most aliphatic compounds are flammable, allowing the use of hydrocarbons as fuel, such as methane in Bunsen burners and as liquified natural gas (LNG), and acetylene in welding.

Aliphatic Acids

Aliphatic acids are the acids of nonaromatic hydrocarbons, such as acetic, propionic, and butyric acids.

Aromaticity

In organic chemistry, aromaticity is a chemical property describing the way in which a conjugated ring of unsaturated bonds, lone pairs, or empty orbitals exhibits a stabilization stronger than would be expected by the stabilization of conjugation alone. The earliest use of the term was in an article by August Wilhelm Hofmann in 1855. There is no general relationship between aromaticity as a chemical property and the olfactory properties of such compounds. Aromaticity can also be considered a manifestation of cyclic delocalization and of resonance. This is usually considered to be because electrons are free to cycle around circular arrangements of atoms that are alternately single- and double-bonded to one another.

These bonds may be seen as a hybrid of a single bond and a double bond, each bond in the ring identical to every other. This commonly seen model of aromatic rings, namely the idea that benzene was formed from a six-membered carbon ring with alternating single and double bonds (cyclohexatriene), was developed by Kekulé.

The model for benzene consists of two resonance forms, which corresponds to the double and single bonds superimposing to give rise to six one-and-a-half bonds. Benzene is a more stable molecule than would be expected without accounting for charge delocalization.

As is standard for resonance diagrams, a double-headed arrow is used to indicate that the two structures are not distinct entities, but merely hypothetical possibilities. Neither is an accurate representation of the *actual* compound, which is best represented by a hybrid (average) of these structures, which can be seen at right. A C=C bond is shorter than a C"C bond, but benzene is perfectly hexagonal—all six carbon-carbon bonds have the same length, intermediate between that of a single and that of a double bond.

A better representation is that of the circular π bond (Armstrong's *inner cycle*), in which the electron density is evenly distributed through a π-bond above and below the ring. This model more correctly represents the location of electron density within the aromatic ring.

Characteristics of Aromatic (aryl) Compounds

An aromatic (or aryl) compound contains a set of covalently bound atoms with specific characteristics:

1. A delocalized conjugated π system, most commonly an arrangement of alternating single and double bonds
2. Coplanar structure, with all the contributing atoms in the same plane

3. Contributing atoms arranged in one or more rings
4. A number of π delocalized electrons that is even, but not a multiple of 4. That is, 4n + 2 number of π electrons, where n=0, 1, 2, 3, and so on. This is known as Hückel's Rule.

Whereas benzene is aromatic (6 electrons, from 3 double bonds), cyclobutadiene is not, since the number of π delocalized electrons is 4, which of course is a multiple of 4. The cyclobutadienide (2–) ion, however, is aromatic (6 electrons). An atom in an aromatic system can have other electrons that are not part of the system, and are therefore ignored for the 4n + 2 rule. In furan, the oxygen atom is sp^2 hybridized. One lone pair is in the π system and the other in the plane of the ring (analogous to C-H bond on the other positions). There are 6 π electrons, so furan is aromatic.

Aromatic molecules typically display enhanced chemical stability, compared to similar non-aromatic molecules. A molecule that can be aromatic will tend to alter its electronic or conformational structure to be in this situation. This extra stability changes the chemistry of the molecule. Aromatic compounds undergo electrophilic aromatic substitution and nucleophilic aromatic substitution reactions, but not electrophilic addition reactions as happens with carbon-carbon double bonds.

Many of the earliest-known examples of aromatic compounds, such as benzene and toluene, have distinctive pleasant smells. This property led to the term "aromatic" for this class of compounds, and hence the term "aromaticity" for the eventually discovered electronic property.

The circulating π electrons in an aromatic molecule produce ring currents that oppose the applied magnetic field in NMR. The NMR signal of protons in the plane of an aromatic ring is shifted substantially further down-field than those on non-aromatic sp^2 carbons. This is an important way of detecting aromaticity. By the same mechanism, the signals of protons located near the ring axis are shifted up-field.

Aromatic molecules are able to interact with each other in so-called π-π stacking: The π systems form two parallel rings overlap in a "face-to-face" orientation. Aromatic molecules are also able to interact with each other in an "edge-to-face" orientation: The slight positive charge of the substituents on the ring atoms of one molecule are attracted to the slight negative charge of the aromatic system on another molecule.

Planar monocyclic molecules containing 4n π electrons are called antiaromatic and are, in general, destabilized. Molecules that could be antiaromatic will tend to alter their electronic or conformational structure to avoid this situation, thereby becoming non-aromatic. For example, cyclooctatetraene (COT) distorts itself out of planarity, breaking π overlap between adjacent double bonds. Relatively recently, cyclobutadiene was discovered to adopt an asymmetric, rectangular configuration in which single and double bonds indeed alternate; there is no resonance and the single bonds are markedly longer than the double bonds, reducing unfavourable p-orbital overlap.

This reduction of symmetry lifts the degeneracy of the two formerly non-bonding molecular orbitals, which by Hund's rule forces the two unpaired electrons into a new, weakly bonding orbital (and also creates a weakly antibonding orbital). Hence, cyclobutadiene is non-aromatic; the strain of the asymmetric configuration outweighs the anti-aromatic destabilization that would afflict the symmetric, square configuration.

Importance of Aromatic Compounds

Aromatic compounds play key roles in the biochemistry of all living things. The four aromatic amino acids histidine, phenylalanine, tryptophan, and tyrosine each serve as one of the 20 basic building-blocks of proteins. Further, all 5 nucleotides (adenine, thymine, cytosine, guanine, and uracil) that make up the sequence of the genetic code in DNA and RNA are aromatic purines or pyrimidines. The molecule heme contains an aromatic system with 22 π electrons. Chlorophyll also has a similar aromatic system.

Aromatic compounds are important in industry. Key aromatic hydrocarbons of commercial interest are benzene, toluene, *ortho*-xylene and *para*-xylene. About 35 million tonnes are produced worldwide every year.

They are extracted from complex mixtures obtained by the refining of oil or by distillation of coal tar, and are used to produce a range of important chemicals and polymers, including styrene, phenol, aniline, polyester and nylon.

Types of Aromatic Compounds

The overwhelming majority of aromatic compounds are compounds of carbon, but they need not be hydrocarbons.

Neutral Homocyclics

Benzene, as well as most other annulenes with the formula C_nH_n where n is an even number, such as cyclotetradecaheptaene.

Heterocyclics

In heterocyclic aromatics (heteroaromats), one or more of the atoms in the aromatic ring is of an element other than carbon. This can lessen the ring's aromaticity, and thus (as in the case of furan) increase its reactivity. Other examples include pyridine, pyrazine, imidazole, pyrazole, oxazole, thiophene, and their benzannulated analogs (benzimidazole, for example).

Polycyclics

Polycyclic aromatic hydrocarbons are molecules containing two or more simple aromatic rings fused together by sharing two neighbouring carbon atoms. Examples are naphthalene, anthracene, and phenanthrene.

Substituted Aromatics

Many chemical compounds are aromatic rings with other functional groups attached. Examples include trinitrotoluene (TNT), acetylsalicylic acid (aspirin), paracetamol, and the nucleotides of DNA.

Atypical Aromatic Compounds

Aromaticity is found in ions as well: the cyclopropenyl cation (2e system), the cyclopentadienyl anion (6e system), the tropylium ion (6e), and the cyclooctatetraene dianion (10e). Aromatic properties have been attributed to non-benzenoid compounds such as tropone. Aromatic properties are tested to the limit in a class of compounds called cyclophanes.

A special case of aromaticity is found in homoaromaticity where conjugation is interrupted by a single sp^3 hybridized carbon atom. When carbon in benzene is replaced by other elements in borabenzene, silabenzene, germanabenzene, stannabenzene, phosphorine or pyrylium salts the aromaticity is still retained. Aromaticity also occurs in compounds that are not carbon-based at all. Inorganic 6-membered-ring compounds analogous to benzene have been synthesized. Hexasilabenzene (Si_6H_6) and borazine ($B_3N_3H_6$) are structurally analogous to benzene, with the carbon atoms replaced by another element or elements. In borazine, the boron and nitrogen atoms alternate around the ring. Quite recently, the aromaticity of planar Si_5^{6-} rings occurring in the Zintl phase $Li_{12}Si_7$ was experimentally evidenced by Li solid state NMR.

Metal aromaticity is believed to exist in certain metal clusters of aluminium. Möbius aromaticity occurs when a cyclic system of molecular orbitals, formed from p_π atomic orbitals and populated in a closed shell by 4n (n is an integer) electrons, is given a single half-twist to correspond to a Möbius strip. A π system with 4n electrons in a flat (non-twisted) ring would be anti-aromatic, and therefore highly unstable, due to the symmetry of the combinations of p atomic orbitals.

By twisting the ring, the symmetry of the system changes and becomes allowed. Because the twist can be left-handed or right-handed, the resulting Möbius aromatics are *dissymmetric* or chiral. Up to now, there is no doubtless proof that a Möbius aromatic molecule was synthesized. Aromatics with two half-twists corresponding to the paradromic topologies, first suggested by Johann Listing, have been proposed by Rzepa in 2005. In carbo-benzene the ring bonds are extended with alkyne and allene groups.

Heterocyclic Compounds

The characteristics of the cyclic hydrocarbons are again altered if heteroatoms are present, which can exist as either substituents attached externally to the ring (exocyclic) or as a member of the ring itself (endocyclic). In the case of the latter, the ring is termed a heterocycle. Pyridine and furan are examples of aromatic heterocycles while piperidine and tetrahydrofuran are the corresponding alicyclic heterocycles. The heteroatom of heterocyclic molecules is generally oxygen, sulfur, or nitrogen, with the latter being particularly common in biochemical systems.

Examples of groups among the heterocyclics are the aniline dyes, the great majority of the compounds discussed in biochemistry such as alkaloids, many compounds related to vitamins, steroids, nucleic acids (e.g. DNA, RNA) and also numerous medicines. Heterocyclics with relatively simple structures are pyrrole (5-membered) and indole (6-membered carbon ring).

Rings can fuse with other rings on an edge to give polycyclic compounds. The purine nucleoside bases are notable polycyclic aromatic heterocycles. Rings can also fuse on a "corner" such that one atom (almost always carbon) has two bonds going to one ring and two to another. Such compounds are termed spiro and are important in a number of natural products.

A heterocyclic compound is a cyclic compound that has atoms of at least two different elements as members of its ring(s). The counterparts of heterocyclic compounds are homocyclic compounds, the rings of which are made of a single element. Although heterocyclic compounds may be inorganic, most contain at least one carbon. Since in organic chemistry non-carbons usually are considered to replace carbon atoms, they are called heteroatoms, meaning 'different from carbon and hydrogen' (rings of heteroatoms of the same element are homocyclic). The IUPAC recommends the Hantzsch-Widman nomenclature for naming heterocyclic compounds.

Heterocyclic chemistry is the branch of chemistry dealing with synthesis, properties, and applications of heterocycles.

Classification based on Electronic Structure

Heterocyclic compounds can be usefully classified based on their electronic structure. The saturated heterocycles behave like the acyclic derivatives. Thus, piperidine and tetrahydrofuran are conventional amines and ethers, with modified steric profiles. Therefore, the study of heterocyclic chemistry focuses especially on unsaturated derivatives, and the preponderance of work and applications involves unstrained 5- and 6-membered rings. Included are pyridine, thiophene, pyrrole, and furan. Another large class of heterocycles are fused to benzene rings, which for pyridine, thiophene, pyrrole, and furan are quinoline, benzothiophene, indole, and benzofuran, respectively. Fusion of two benzene rings gives rise to a third large family of compounds, respectively the acridine, dibenzothiophene, carbazole, and dibenzofuran. The unsaturated rings can be classified according to the participation of the heteroatom in the pi-system.

Fused rings

Heterocyclic rings systems that are formally derived by fusion with other rings, either carbocyclic or heterocyclic, have a variety of common and systematic names. For example, with the benzo-fused unsaturated nitrogen heterocycles, pyrrole provides indole or isoindole depending on the orientation. The pyridine analog is quinoline or isoquinoline.

For azepine, benzazepine is the preferred name. Likewise, the compounds with two benzene rings fused to the central heterocycle are carbazole, acridine, and dibenzoazepine.

History of Heterocyclic Chemistry

The history of heterocyclic chemistry began in the 1800s, in step with the development of organic chemistry. Some noteworthy developments:

1818: Brugnatelli isolates alloxan from uric acid

1832: Dobereiner produces furfural (a furan) by treating starch with sulfuric acid

1834: Runge obtains pyrrole ("fiery oil") by dry distillation of bones

1906: Friedlander synthesizes indigo dye, allowing synthetic chemistry to displace a large agricultural industry

1936: Treibs isolates chlorophyl derivatives from crude oil, explaining the biological origin of petroleum.

1951: Chargaff's rules are described, highlighting the role of heterocyclic compounds (purines and pyrimidines) in the genetic code.

Commercial Exploitation

Leading companies with a vast amount of patents related to heterocyclic compounds are Bayer, Merck, Ciba-Geigy, Pfizer, Eli Lily, BASF, Hoffmann La Roche, ER Sqibb, Warner Lambert and Hoechst.

Polymers

One important property of carbon is that it readily forms chains, or networks, that are linked by carbon-carbon (carbon to carbon) bonds. The linking process is called polymerization, while the chains, or networks, are called polymers. The source compound is called a monomer. Two main groups of polymers exist: synthetic polymers and biopolymers. Synthetic polymers are artificially manufactured, and are commonly referred to as industrial polymers. Biopolymers occur within a respectfully natural environment, or without human intervention.

Since the invention of the first synthetic polymer product, bakelite, synthetic polymer products have frequently been invented. Common synthetic organic polymers are polyethylene (polythene), polypropylene, nylon, teflon (PTFE), polystyrene, polyesters, polymethylmethacrylate (called perspex and plexiglas), and polyvinylchloride (PVC). Both synthetic and natural rubber are polymers. Varieties of each synthetic polymer product may exist, for purposes of a specific use. Changing the conditions of polymerization alters the chemical composition of the product and its properties.

These alterations include the chain length, or branching, or the tacticity. With a single monomer as a start, the product is a homopolymer. Secondary component(s) may be added to create a heteropolymer (co-polymer) and the degree of clustering of the different components can also be controlled.

Physical characteristics, such as hardness, density, mechanical or tensile strength, abrasion resistance, heat resistance, transparency, colour, etc. will depend on the final composition.

Biomolecules

Biomolecular chemistry is a major category within organic chemistry which is frequently studied by biochemists. Many complex multi-functional group molecules are important in living organisms. Some are long-chain biopolymers, and these include peptides, DNA, RNA and the polysaccharides such as starches in animals and celluloses in plants.

The other main classes are amino acids (monomer building blocks of peptides and proteins), carbohydrates (which includes the polysaccharides), the nucleic acids (which include DNA and RNA as polymers), and the lipids.

In addition, animal biochemistry contains many small molecule intermediates which assist in energy production through the Krebs cycle, and produces isoprene, the most common hydrocarbon in animals. Isoprenes in animals form the important steroid structural (cholesterol) and steroid hormone compounds; and in plants form terpenes, terpenoids, some alkaloids, and a unique set of hydrocarbons called biopolymer polyisoprenoids present in latex sap, which is the basis for making rubber.

Peptide Synthesis

In organic chemistry, peptide synthesis is the production of peptides, which are organic compounds in which multiple amino acids are linked via amide bonds which are also known as peptide bonds. The biological process of producing long peptides (proteins) is known as protein biosynthesis.

Peptides are synthesized by coupling the carboxyl group or C-terminus of one amino acid to the amino group or N-terminus of another. Due to the possibility of unintended reactions, protecting groups are usually necessary. Chemical peptide synthesis starts at the C-terminal end of the peptide and ends at the N-terminus. This is the opposite of protein biosynthesis, which starts at the N-terminal end.

Liquid-Phase Synthesis

Liquid-phase peptide synthesis is a classical approach to peptide synthesis. It has been replaced in most labs by solid-phase synthesis. However, it retains usefulness in large-scale production of peptides for industrial purposes.

Solid-Phase Synthesis

Solid-phase peptide synthesis (SPPS), pioneered by Robert Bruce Merrifield, resulted in a paradigm shift within the peptide synthesis community. It is now the accepted method for creating peptides and proteins in the lab in a synthetic manner. SPPS allows the synthesis of natural peptides which are difficult to express in bacteria, the incorporation of unnatural amino acids, peptide/protein backbone modification, and the synthesis of D-proteins, which consist of D-amino acids. Small solid beads, insoluble yet porous, are treated with functional units ('linkers') on which peptide chains can be built. The peptide will remain covalently attached to the bead until cleaved from it by a reagent such as anhydrous hydrogen fluoride or trifluoroacetic acid. The peptide is thus 'immobilized' on the solid-phase and can be retained during a filtration process, whereas liquid-phase reagents and by-products of synthesis are flushed away.

The general principle of SPPS is one of repeated cycles of coupling-wash-deprotection-wash. The free N-terminal amine of a solid-phase attached peptide is coupled to a single N-protected amino acid unit.

This unit is then deprotected, revealing a new N-terminal amine to which a further amino acid may be attached. The superiority of this technique partially lies in the ability to perform wash cycles after each reaction, removing excess reagent with all of the growing peptide of interest remaining covalently attached to the insoluble resin.

The overwhelmingly important consideration is to generate extremely high yield in each step. For example, if each coupling step were to have 99% yield, a 26-amino acid peptide would be synthesized in 77% final yield (assuming 100% yield in each deprotection); if each step were 95%, it would be synthesized in 25% yield. Thus each amino acid is added in major excess (2~10x) and coupling amino acids together is highly optimized by a series of well-characterised agents.

There are two majorly used forms of SPPS — Fmoc and Boc. Unlike ribosome protein synthesis, solid-phase peptide synthesis proceeds in a C-terminal to N-terminal fashion. The N-termini of amino acid monomers is protected by either of these two groups and added onto a deprotected amino acid chain. Automated synthesizers are available for both techniques, though many research groups continue to perform SPPS manually.

SPPS is limited by yields, and typically peptides and proteins in the range of 70 amino acids are pushing the limits of synthetic accessibility. Synthetic difficulty also is sequence dependent; typically amyloid peptides and proteins are difficult to make. Longer lengths can be accessed by using native chemical ligation to couple two peptides together with quantitative yields.

Since its introduction over 40 years ago, SPPS has been significantly optimized. First, the resins themselves have been optimized. Furthermore, the 'linkers' between the C-terminal amino acid and polystyrene resin have improved attachment and cleavage to the point of mostly quantitative yields.

The evolution of side chain protecting groups has limited the frequency of unwanted side reactions. In addition, the evolution of new activating groups on the carboxyl group of the incoming amino acid have improved coupling and decreased epimerization. Finally, the process itself has been optimized.

In Merrifield's initial report, the deprotection of the α-amino group resulted in the formation of a peptide-resin salt, which required neutralization with base prior to coupling.

The time between neutralization of the amino group and coupling of the next amino acid allowed for aggregation of peptides, primarily through the formation of secondary structures, and adversely affected coupling.

The Kent group showed that concomitant neutralization of the α-amino group and coupling of the next amino acid led to improved coupling. Each of these improvements has helped SPPS become the robust technique that it is today.

BOP SPPS

The use of BOP reagent was first described by Castro et al. in 1975.

Solid Supports

The name solid support implies that reactions are carried out on the surface of the support, but this is not the case. Reactions also occur within these particles, and thus the term "solid support" better describes the insolubility of the polymer. The physical properties of the solid support, and the applications to which it can be utilised, vary with the material from which the support is constructed, the amount of cross-linking, as well as the linker and handle being used.

Most scientists in the field believe that supports should have the minimum amount of cross-linking to confer stability. This should result in a well-solvated system where solid-phase peptide synthesis can be carried out. Nonetheless, the characteristics of an efficient solid support include:

1. It must be physically stable and permit the rapid filtration of liquids, such as excess reagents
2. It must be inert to all reagents and solvents used during SPPS

3. It must swell extensively in the solvents used to allow for penetration of the reagents
4. It must allow for the attachment of the first amino acid

There are four primary types of solid supports:

1. ***Gel-type supports:*** These are highly solvated polymers with an equal distribution of functional groups. This type of support is the most common, and includes:
   * *Polystyrene:* Styrene cross-linked with 1-2% divinylbenzene
   * *Polyacrylamide:* A hydrophilic alternative to polystyrene
   * *Polyethylene glycol (PEG):* PEG-Polystyrene (PEG-PS) is more stable than polystyrene and spaces the site of synthesis from the polymer backbone
   * *PEG-based supports:* Composed of a PEG-polypropylene glycol network or PEG with polyamide or polystyrene
2. ***Surface-type supports:*** Many materials have been developed for surface functionalization, including controlled pore glass, cellulose fibres, and highly cross-linked polystyrene.
3. ***Composites:*** Gel-type polymers supported by rigid matrices.

Polyamide Resin

Polyamide resin is also a useful and versatile resin. It seems to swell much more than polystyrene, in which case it may not be suitable for some automated synthesizers, if the wells are too small.

PEG Hybrid Polystyrene Resin

An example of this type of resin is the Tentagel resin. The base resin is polystyrene onto which is attached long chains (Mw ca. 3000 Da) of polyethylene glycol (PEG; also known as polyethylene oxide).

Synthesis is caried out on the distal end of the PEG spacer making it suited for long and difficult peptides. In addition it is also attractive for the synthesis of combinatorial Peptide libraries and on resin screening experiments. It does not expand much during synthesis making it a preferred resin for robotic peptide synthesis.

PEG Based Resin

ChemMatrix(R) is a new type of resin which is based on PEG that is crosslinked. ChemMatrix(R) has claimed a high chemical and thermal stability (is compatible with Microwave synthesis) and has shown higher degrees of swellings in acetonitrile, dichloromethane, DMF, N-methylpyrrolidone, TFA and water compared to the polystyrene based resins. ChemMatrix has shown significant improvements to the synthesis of hydrophobic sequences. ChemMatrix may be useful for the synthesis of difficult and long peptides.

Improvements to solid supports used for peptide synthesis enhance their ability to withstand the repeated use of TFA during the deprotection step of SPPS. Furthermore, different resins allow for different functional groups at the C-terminus. The oxymethylphenylacetamidomethyl (PAM) resin results in the conventional C-terminal carboxylic acid. On the other hand, the paramethylbenzhydrylamine (pMBHA) resin yields a C-terminal amide, which is useful in mimicking the interior of a protein.

Along with the development of Fmoc SPPS, different resins have also been created to be removed by TFA. Similar to the Boc strategy, two primary resins are used, based on whether a C-terminal carboxylic acid or amide is desired. The Wang resin is the most commonly used resin for peptides with C-terminal carboxylic acids. If a C-terminal amide is desired, the Rink amide resin is used.

Protecting Groups

Amino acids have reactive moieties at the N- and C-termini, which facilitates amino acid coupling during synthesis.

Many amino acids also have reactive side chain functional groups, which can interact with free termini or other side chain groups during synthesis and peptide elongation and negatively influence yield and purity.

To facilitate proper amino acid synthesis with minimal side chain reactivity, chemical groups have been developed to bind to specific amino acid functional groups and block, or protect, the functional group from nonspecific reactions. These protecting groups, while vast in nature, can be separated into three groups, as follows:

- N-terminal protecting groups
- C-terminal protecting groups (mostly used in liquid-phase synthesis)
- Side chain protecting groups

Purified, individual amino acids are reacted with these protecting groups prior to synthesis and then selectively removed during specific steps of peptide synthesis.

N-Terminal Protecting Groups

Amino acids are added to excess to ensure complete coupling during each synthesis step, and without N-terminal protection, polymerization of unprotected amino acids could occur, resulting in low peptide yield or synthesis failure.

N-terminal protection requires an additional step of removing the protecting group, termed deprotection, prior to the coupling step, creating a repeating design flow as follows:

- Protecting group is removed from the trailing amino acids in a deprotection reaction
- Deprotection reagents are washed away to provide a clean coupling environment
- Protected amino acids dissolved in a solvent such as dimethylformamide (DMF) combined with coupling reagents are pumped through the synthesis column
- Coupling reagents are washed away to provide clean deprotection environment

Currently, two protecting groups (t-Boc, Fmoc) are commonly used in solid-phase peptide synthesis. Their liability is caused by the carbamate group which readily releases CO_2 for an irreversible decoupling step.

Protection and Deprotection of Amines

The Boc group can be added to the amine under aqueous conditions using di-*tert*-butyl dicarbonate in the presence of a base such as sodium bicarbonate. Protection of the amine can also be accomplished in acetonitrile solution using 4-dimethylaminopyridine (DMAP) as the base. Removal of the *t*-

BOC in amino acids can be accomplished with strong acids such as trifluoroacetic acid neat or in dichloromethane, or with HCl in methanol. A complication may be the tendency of the t-butyl cation intermediate to alkylate other nucleophiles; scavengers such as anisole or thioanisole may be used. Selective cleavage of the N-Boc group in the presence of other protecting groups is possible when using $AlCl_3$. The use of triethylsilane as a carbocation scavenger in the presence of trifluoroacetic acid in dichloromethane has been shown to lead to increased yields, decreased reaction times, simple work-up and improved selectivity for the deprotection of t-butyl ester and t-butoxycarbonyl sites in protected amino-acids and peptides in the presence of other acid-sensitive protecting groups such as the benzyloxycarbonyl, 9-fluorenylmethoxycarbonyl, O- and S-benzyl and t-butylthio groups.

Alloc Protecting Group

The allyloxycarbonyl (alloc) protecting group is often used to protect a carboxylic acid, hydroxyl, or amino group when an orthogonal deprotection scheme is required. It is sometimes used when conducting on-resin cyclic peptide formation, where the peptide is linked to the resin by a side-chain functional group.

The alloc group can be removed using tetrakis (triphenylphosphine) palladium (0) along with a 37:2:1 mixture of methylene chloride, acetic acid, and N-Methylmorpholine (NMM) for 2 hours. The resin must then be carefully washed 0.5% DIPEA in DMF, 3x10 ml of 0.5% sodium diethylthiocarbamate in DMF, and then 5x10 ml of 1:1 DCM:DMF.

Lithographic Protecting Groups

For special applications like protein microarrays lithographic protecting groups are used. Those groups can be removed through exposure to light.

Side Chain Protecting Groups

Amino acid side chains represent a broad range of functional groups and are sites of nonspecific reactivity during peptide synthesis. Because of this, many different protecting groups are required that are usually based on the benzyl (Bzl) or tert-butyl (tBu) group. The specific protecting groups used during the synthesis of a given peptide vary depending on the peptide sequence and the type of N-terminal protection used. Side chain protecting groups are known as permanent or semipermanent protecting groups, because they can withstand the multiple cycles of chemical treatment during synthesis and are only removed during treatment with strong acids after peptide synthesis is completed.

Protection Schemes

Because multiple protecting groups are normally used during peptide synthesis, these groups must be compatible to allow deprotection of distinct protecting groups while not affecting other protecting groups. Protecting schemes are therefore established to match protecting groups so that deprotection of one protecting group does not affect the binding of the other groups.

Because N-terminal deprotection occurs continuously during peptide synthesis, protecting schemes have been established in which the different types of side chain protecting groups (Bzl or tBu) are matched to either Boc or Fmoc, respectively, for optimized deprotection. These protecting schemes also incorporate each of the steps of synthesis and cleavage.

Activating Groups

For coupling the peptides the carboxyl group is usually activated. This is important for speeding up the reaction. There are two main types of activating groups: carbodiimides and triazolols. However the use of pentafluorophenyl esters (FDPP, PFPOH) and BOP-Cl are useful for cyclising peptides.

Carbodiimides

These activating agents were first developed. Most common are dicyclohexylcarbodiimide (DCC) and diisopropylcarbodiimide (DIC). Reaction with a carboxylic acid yields a highly reactive O-acyl-urea. During artificial protein synthesis (such as Fmoc solid-state synthesizers), the C-terminus is often used as the attachment site on which the amino acid monomers are added. To enhance the electrophilicity of carboxylate group, the negatively charged oxygen must first be “activated” into a better leaving group. DCC is used for this purpose. The negatively charged oxygen will act as a nucleophile, attacking the central carbon in DCC. DCC is temporarily attached to the former carboxylate group (which is now an ester group), making nucleophilic attack by an amino group (on the attaching amino acid) to the former C-terminus (carbonyl group) more efficient. The problem with carbodiimides is that they are too reactive and that they can therefore cause racemization of the amino acid.

Regioselective Disulfide Formation

The formation of multiple native disulfides remains one of the primary challenges of native peptide synthesis by solid-phase methods. Random chain combination typically results in several products with nonnative disulfide bonds. Stepwise formation of disulfide bonds is typically the preferred method, and performed with thiol protecting groups (PGs). Different thiol PGs provide multiple dimensions of orthogonal protection.

These orthogonally-protected cysteines are incorporated during the solid-phase synthesis of the peptide. Successive removal of these PGs to allow for selective exposure of free thiol groups, leads to disulfide formation in a stepwise manner.

The order of removal of these PGs must be considered so that only one group is removed at a time. Using this method, Kiso *et al.* reported the first total synthesis of insulin by this method in 1993. The thiol PGs must possess multiple characteristics. First, the PG must be reversible with conditions that do not affect the unprotected side chains. Second, the protecting group must be able to withstand the conditions of solid-phase synthesis. Third, the configuration of the removal of the thiol protecting group must be such that it leaves intact other thiol PGs, if orthogonal protection is desired. That is, the removal of PG A should not affect PG B. Some of the thiol PGs commonly used includes the acetamidomethyl (Acm), tert-butyl (But), 3-nitro-2-pyridine sulfenyl (NPYS), 2-pyridine-sulfenyl (Pyr), and triphenylmethyl (Trt) groups. Importantly, the NPYS group can replace the Acm PG to yield an activated thiol.

In the stepwise formation of disulfides to synthesize insulin by Kiso *et al.*, the authors synthesize the A-chain with following protection: CysA6(But); CysA7(Acm); CysA11(But). Thus,

CysA20 is unprotected. Synthesis of the B-chain is performed with the following protection: CysB7(Acm) CysB19(Pyr). The first disulfide bond, CysA20-CysB19, was formed by mixing the two chains in 8 M urea, pH 8 (RT) for 50 min. The second disulfide bond, CysA7-CysB7, was formed by treatment with iodine in aqueous acetic acid to remove the Acm groups. The third disulfide, the intramolecular CysA6-CysA11, was formed by the removal of the But groups by methyltrichlorosilane with diphenyl sulfoxide in TFA. Importantly, formation of the first disulfide in 8 M urea, pH 8 does not affect the other PGs, namely Acm and But groups. Likewise, formation of the second disulfide bond with iodine in aqueous acetic acid does not affect the But groups.

Important to the discussion of disulfide bond formation is the order in which disulfides are formed. From a logical standpoint, the order in which the thiol groups are exposed to form disulfides should be of little consequence, since the other cysteines are protected. Practically, however, the order in which disulfides are formed can have a significant effect on yields. This may be because the formation of the CysA20-CysB19 disulfide may place the thiol group of CysB7 in close proximity with both CysA6 and CysA7, leading to multiple disulfide products. This is one manifestation of the reality that solid-phase peptide synthesis is as much art as it is science.

Synthesizing Long Peptides

Stepwise elongation, in which the amino acids are connected step-by-step in turn, is ideal for small peptides containing between 2 and 100 amino acid residues. Another method is fragment condensation, in which peptide fragments are coupled. Although the former can elongate the peptide chain without racemization, the yield drops if only it is used in the creation of long or highly polar peptides. Fragment condensation is better than stepwise elongation for synthesizing sophisticated long peptides, but its use must be restricted in order to protect against racemization. Fragment condensation is also undesirable since the coupled fragment must be in gross excess, which may be a limitation depending on the length of the fragment.

A new development for producing longer peptide chains is chemical ligation: Unprotected peptide chains react chemoselectively in aqueous solution. A first kinetically controlled product rearranges to form the amide bond. The most common form of native chemical ligation uses a peptide thioester that reacts with a terminal cysteine residue. In order to optimize synthesis of long peptides, Zealand Pharma (located in Denmark in Medicon Valley) invented a method for converting a difficult peptide sequence into an easy peptide sequence. The new technology, called SIP-technology, uses "structure-inducing probes" (SIP) to facilitate the synthesis of long peptides. The SIP-technology is a small pre-sequence peptide sequence (e.g. Lysine (Lysn); Glutamic Acid (Glun); (LysGlu)n) that is incorporated at the C-terminus of subsequent resin bound peptide to induce an alpha-helix-like structure in the peptide. The SIP technology constrains the parent peptide into a more ordered conformation using intramolecular hydrogen bonds. This allows the peptide structure to stabilize, and the utilised hydrogen bonds reduce the likelihood of aggregation and degradation by enzymes. In this way, the SIP technology is designed to optimize peptide synthesis, increase biological half-life, improve peptide stability and inhibit enzymatic degradation without altering pharmacological activity or profile of action.

Microwave Assisted Peptide Synthesis

Although microwave irradiation has been around since the late 1940s, it was not until 1986 that microwave energy was used in organic chemistry. During the end of the 1980s and 1990s, microwave energy was an obvious source for completing chemical reactions in minutes that would otherwise take several hours to days.

Through several technical improvements at the end of the 1990s and beginning of the 2000s, microwave synthesizers have been designed to provide both low and high energy pockets of microwave energy so that the temperature of the reaction mixture could be controlled. The microwave energy used in peptide synthesis is of a single frequency providing maximum penetration depth of the sample which is in contrast to conventional kitchen microwaves.

In peptide synthesis, microwave irradiation has been used to complete long peptide sequences with high degrees of yield and low degrees of racemization. Microwave irradiation during the coupling of amino acids to a growing polypeptide chain is not only catalysed through the increase in temperature, but also due to the alternating electromagnetic radiation to which the polar backbone of the polypeptide continuously aligns to. Due to this phenomenon, the microwave energy can prevent aggregation and thus increases yields of the final peptide product.

There is however no clear evidence that microwave is better than simple heating and some peptide laboratories regard microwave just as a convenient method for rapid heating of the peptidyl resin. Heating to above 50-55 degrees Celsius also prevents aggregation and accelerates the coupling.

Despite the main advantages of microwave irradiation of peptide synthesis, the main disadvantage is the racemization which may occur with the coupling of cysteine and histidine. A typical coupling reaction with these amino acids is performed at lower temperatures than the other 18 natural amino acids. A number of peptides do not survive microwave synthesis or heating in general. One of the more serious side effects is dehydration (loss of water) which for certain peptides can be almost quantitative like pancreatic polypeptide (PP). This side effect is also seen by simple heating without the use of microwave. Microwave chemistry is the science of applying microwave irradiation to chemical reactions. Microwaves act as high frequency electric fields and will generally heat any material containing mobile electric charges, such as polar molecules in a solvent or conducting ions in a solid. Polar solvents are heated as their component molecules are forced to rotate with the field and lose energy in collisions. Semiconducting and conducting samples heat when ions or electrons within them form an electric current and energy is lost due to the electrical resistance of the material.

Microwave heating in the laboratory began to gain wide acceptance following papers in 1986, although the use of microwave heating in chemical modification can be traced back to the 1950s. Although occasionally known by such acronyms as 'MEC' (Microwave-Enhanced Chemistry) or *MORE synthesis* (Microwave-organic Reaction Enhancement), these acronyms have had little acceptance outside a small number of groups.

Conventional heating usually involves the use of a furnace or oil bath, which heats the walls of the reactor by convection or conduction. The core of the sample takes much longer to achieve the target temperature, e.g. when heating a large sample of ceramic bricks.

Microwave heating is able to heat the target compounds without heating the entire furnace or oil bath, which saves time and energy.

It is also able to heat sufficiently thin objects throughout their volume (instead of through its outer surface), in theory producing more uniform heating. However, due to the design of most microwave ovens and to uneven absorption by the object being heated, the microwave field is usually non-uniform and localized superheating occurs.

Different compounds convert microwave radiation to heat by different amounts. This selectivity allows some parts of the object being heated to heat more quickly or more slowly than others (particularly the reaction vessel).

Microwave heating can have certain benefits over conventional ovens:

- Reaction rate acceleration
- Milder reaction conditions
- Higher chemical yield
- Lower energy usage
- Different reaction selectivities

Selective Heating

A heterogeneous system (comprising different substances or different phases) may be anisotropic if the loss tangents of the components are considered. As a result, it can be expected that the microwave field energy will be converted to heat by different amounts in different parts of the system.

This inhomogeneous energy dissipation means selective heating of different parts of the material is possible, and may lead to temperature gradients between them. Nevertheless, the presence of zones with a higher temperature than others (called hot spots) must be subjected to the heat transfer processes between domains. Where the rate of heat conduction is high between system domains, hot spots would have no long-term existence as the components rapidly reach thermal equilibrium. In a system where the heat transfer is slow, it would be possible to have the presence of a steady state hot spot that may enhance the rate of the chemical reaction within that hot zone.

On this basis, many early papers in microwave chemistry postulated the possibility of exciting specific molecules, or functional groups within molecules. However, the time within which thermal energy is repartitioned from such moieties is much shorter than the period of a microwave wave, thus precluding the presence of such 'molecular hot spots' under ordinary laboratory conditions. The oscillations produced by the radiation in these target molecules would be instantaneously transferred by collisions with the adjacent molecules, reaching at the same moment the thermal equilibrium. Processes with solid phases behave somewhat differently. In this case much higher heat transfer resistances are involved, and the possibility of the stationary presence of hot-spots should be contemplated. A differentiation between two kinds of hot spots has been noted in the literature, although the distinction is considered by many to be arbitrary. *Macroscopic hot spots* were considered to comprise all large non-isothermal volumes that can be detected and measured by use of optical pyrometres (optical fibre or IR). By these means it is possible to visualise thermal inhomogeneities within solid phases under microwave irradiation. *Microscopic hot spots* are non-isothermal regions that exist at the micro- or nanoscale (e.g. supported metal nanoparticles inside a catalyst pellet) or in the molecular scale (e.g. a polar group on a catalyst structure). The distinction has no serious significance, however, as microscopic hotspots such as those proposed to explain catalyst behaviour in several gas-phase catalytic reactions have been demonstrated by post-mortem methods and in-situ methods. Some theoretical and experimental approaches have been published towards the clarification of the hot spot effect in heterogeneous catalysts.

A different specific application in synthetic chemistry is in the microwave heating of a binary system comprising a polar solvent and a non-polar solvent obtain different temperatures. Applied in a phase transfer reaction a water phase reaches a temperature of 100°C while a chloroform phase would retain a temperature of 50°C, providing the extraction as well of the reactants from one phase to the other. Microwave chemistry is particularly effective in dry media reactions.

The phrase microwave effect is a term that is applied to a range of observations in microwave chemistry. There are two general classes of microwave effects:

- Specific microwave effects.
- Non-thermal microwave effects.

A recent review has proposed this definition and examples of microwave effects in organic chemistry have been summarised. Specific microwave effects are those effects that cannot be (easily) emulated through conventional heating methods. Examples include: (i) selective heating of specific reaction components, (ii) rapid heating rates and temperature gradients, (iii) the elimination of wall effects, and (iv) the superheating of solvents. Microwave-specific effects tend not to be controversial and invoke "conventional" explanations (i.e. kinetic effects) for the observed effects. Non-thermal microwave effects have been proposed in order to explain unusual observations in microwave chemistry. As the name suggests, the effects are supposed not to require the transfer of microwave energy into thermal energy. Instead, the microwave energy itself directly couples to energy modes within the molecule or lattice. Non-thermal effects in liquids are almost certainly non-existent, as the time for energy redistribution between molecules in a liquid is much less than the period of a microwave oscillation.

Non-thermal effects in solids are still part of an ongoing debate. It is likely that, through focusing of electric fields at particle interfaces, microwaves cause plasma formation and enhance diffusion in solids via second-order effects. As a result, they may enhance solid-state sintering and diffusion processes. Debates are still raging (January 2006) about non-thermal effects of microwaves that have been reported in solid-state phase transitions.

Non-thermal microwave effects have been posited in order to explain unusual observations in microwave chemistry. As the name suggests, the effects are supposed not to require the transfer of microwave energy into thermal energy. Instead, the microwave energy itself directly couples to energy modes within the molecule or lattice. Non-thermal effects in liquids are almost certainly non-existent, as the time for energy redistribution between molecules in a liquid is much less than the period of a microwave oscillation. A recent review has illustrated this in application to organic chemistry, though clearly supports the existence of non-thermal effects. It has been shown that such non-thermal effects exist in the reaction of O + HCl(DCl)->OH(OD)+Cl in the gas phase and the authors suggest that some mechanisms may also be present in the condensed phase. Non-thermal effects in solids are still part of an ongoing debate. It is likely that, through focusing of electric fields at particle interfaces, microwaves cause plasma formation and enhance diffusion in solids via second-order effects. As a result, they may enhance solid-state sintering processes. Debates are still raging (January 2006) about non-thermal effects of microwaves that have been reported in solid-state phase transitions.

Chapter 2

Physical and Chemical Properties

Alcohols have an odour that is often described as "biting" and as "hanging" in the nasal passages. Ethanol has a slightly sweeter (or more fruit-like) odour than the other alcohols.

In general, the hydroxyl group makes the alcohol molecule polar. Those groups can form hydrogen bonds to one another and to other compounds (except in certain large molecules where the hydroxyl is protected by steric hindrance of adjacent groups). This hydrogen bonding means that alcohols can be used as protic solvents. Two opposing solubility trends in alcohols are: the tendency of the polar OH to promote solubility in water, and the tendency of the carbon chain to resist it. Thus, methanol, ethanol, and propanol are miscible in water because the hydroxyl group wins out over the short carbon chain. Butanol, with a four-carbon chain, is moderately soluble because of a balance between the two trends. Alcohols of five or more carbons (pentanol and higher) are effectively insoluble in water because of the hydrocarbon chain's dominance. All simple alcohols are miscible in organic solvents. Because of hydrogen bonding, alcohols tend to have higher boiling points than comparable hydrocarbons and ethers. The boiling point of the alcohol ethanol is 78.29 °C, compared to 69 °C for the hydrocarbon hexane (a common constituent of gasoline), and 34.6 °C for diethyl ether.

Alcohols, like water, can show either acidic or basic properties at the -OH group. With a pK_a of around 16-19, they are, in general, slightly weaker acids than water, but they are still able to react with strong bases such as sodium hydride or reactive metals such as sodium. The salts that result are called alkoxides, with the general formula $RO^- M^+$. They can react to form ester compounds, and they can (if activated first) undergo nucleophilic substitution reactions. The lone pairs of electrons on the oxygen of the hydroxyl group also makes alcohols nucleophiles.

As one moves from primary to secondary to tertiary alcohols with the same backbone, the hydrogen bond strength, the boiling point, and the acidity typically decrease.

Alcohol has a long history of several uses worldwide. It is found in beverages for adults, as fuel, and also has many scientific, medical, and industrial uses.

The term alcohol-free is often used to describe a product that does not contain alcohol. Some consumers of some commercially prepared products may view alcohol as an undesirable ingredient, particularly in products intended for children.

Alcoholic Beverages

Alcoholic beverages, typically containing 5% to 40% ethanol by volume, have been produced and consumed by humans since pre-historic times.

Antifreeze

A 50% v/v (by volume) solution of ethylene glycol in water is commonly used as an antifreeze.

Antiseptics

Ethanol can be used as an antiseptic to disinfect the skin before injections are given, often along with iodine.

Ethanol-based soaps are becoming common in restaurants and are convenient because they do not require drying due to the volatility of the compound. Alcohol based gels have become common as hand sanitizers.

Fuels

Some alcohols, mainly ethanol and methanol, can be used as an alcohol fuel. Fuel performance can be increased in forced induction internal combustion engines by injecting alcohol into the air intake after the turbocharger or supercharger has pressurized the air. This cools the pressurized air, providing a denser air charge, which allows for more fuel, and therefore more power.

Preservative

Alcohol is often used as a preservative for specimens in the fields of science and medicine.

Solvents

Alcohols have applications in industry and science as reagents or solvents. Because of its relatively low toxicity compared with other alcohols and ability to dissolve non-polar substances, ethanol can be used as a solvent in medical drugs, perfumes, and vegetable essences such as vanilla. In organic synthesis, alcohols serve as versatile intermediates.

Production

In industry, alcohols are produced in several ways:

- By fermentation using glucose produced from sugar from the hydrolysis of starch, in the presence of yeast and temperature of less than 37 °C to produce ethanol. For instance, such a process might proceed by the conversion of sucrose by the enzyme invertase into glucose and fructose, then the conversion of glucose by the enzyme zymase into ethanol (and carbon dioxide).
- By direct hydration using ethylene (ethylene hydration) or other alkenes from cracking of fractions of distilled crude oil.

Endogenous

Several of the benign bacteria in the intestine use fermentation as a form of anaerobic metabolism. This metabolic reaction produces ethanol as a waste product, just like aerobic respiration produces carbon dioxide and water. Thus, human bodies contain some quantity of alcohol endogenously produced by these bacteria.

Substitution

Primary alkyl halides react with aqueous NaOH or KOH mainly to primary alcohols in nucleophilic aliphatic substitution. (Secondary and especially tertiary alkyl halides will give the elimination (alkene) product instead). Grignard reagents react with carbonyl groups to secondary and tertiary alcohols. Related reactions are the Barbier reaction and the Nozaki-Hiyama reaction.

Reduction

Aldehydes or ketones are reduced with sodium borohydride or lithium aluminium hydride (after an acidic workup).

Another reduction by aluminiumisopropylates is the Meerwein-Ponndorf-Verley reduction.

Noyori asymmetric hydrogenation is the asymmetric reduction of β-keto-esters.

Deprotonation

Alcohols can behave as weak acids, undergoing deprotonation. The deprotonation reaction to produce an alkoxide salt is performed either with a strong base such as sodium hydride or *n*-butyllithium or with sodium or potassium metal.

$$2\ R\text{-}OH + 2\ NaH \rightarrow 2\ R\text{-}O^{-}Na^{+} + 2H_2-$$

$$2\ R\text{-}OH + 2\ Na \rightarrow 2\ R\text{-}O^{"}Na^{+} + H_2$$

$$2\ CH_3CH_2\text{-}OH + 2\ Na \rightarrow 2\ CH_3\text{-}CH_2\text{-}O^{"+} + H_2-$$

Water is similar in pK_a to many alcohols, so with sodium hydroxide there is an equilibrium set-up, which usually lies to the left:

R-OH + NaOH Ì! R-O⁻Na⁺ + H_2O (equilibrium to the left)

It should be noted, however, that the bases used to deprotonate alcohols are strong themselves. The bases used and the alkoxides created are both highly moisture-sensitive chemical reagents.

The acidity of alcohols is also affected by the overall stability of the alkoxide ion. Electron-withdrawing groups attached to the carbon containing the hydroxyl group will serve to stabilize the alkoxide when formed, thus resulting in greater acidity.

On the other hand, the presence of electron-donating group will result in a less stable alkoxide ion formed.

This will result in a scenario whereby the unstable alkoxide ion formed will tend to accept a proton to reform the original alcohol. With alkyl halides alkoxides give rise to ethers in the Williamson ether synthesis.

Oxidation to Aldehydes

Reagents useful for the transformation of primary alcohols to aldehydes are normally also suitable for the oxidation of secondary alcohols to ketones. These include:

- Chromium-based reagents, such as Collins reagent ($CrO_3 \cdot Py_2$), PDC or PCC.
- Activated DMSO, resulting from reaction of DMSO with electrophiles, such as oxalyl chloride (Swern oxidation), a carbodiimide (Pfitzner-Moffatt oxidation) or the complex $SO_3 \cdot Py$ (Parikh-Doering oxidation).
- Hypervalent iodine compounds, such as Dess-Martin periodinane or 2-Iodoxybenzoic acid.
- Catalytic TPAP in presence of excess of NMO (Ley oxidation).
- Catalytic TEMPO in presence of excess bleach (NaOCl) (Anelli's oxidation).

Allylic and benzylic alcohols can be oxidized in presence of other alcohols using certain selective oxidants such as manganese dioxide (MnO_2).

Potassium Dichromate

A secondary alcohol can be oxidised into a ketone using acidified potassium dichromate and heating under reflux. The orange-red dichromate ion, $Cr_2O_7^{2-}$, is reduced to the green Cr^{3+} ion. This reaction was once used in an alcohol breath test.

Dess-Martin Oxidation

The Dess-Martin Reagent is a mild oxidant for the conversion of alcohols to aldehydes or ketones. The reaction is performed under standard conditions, at room temperature, with either dichloromethane or trichloromethane. The reaction takes between half an hour and two hours to complete. The products are then separated from the byproduct, an iodide compound.

Advantages of Use

An advantage of the Oppenauer oxidation is its use of relatively inexpensive and non-toxic reagents. Reaction conditions are mild and gentle since the substrates are generally heated in acetone/benzene mixtures. Another advantage of the Oppenauer oxidation which makes it unique to other oxidation methods such as Pyridinium chlorochromate (PCC) and Dess-Martin periodinane is that secondary alcohols are oxidized much faster than primary alcohols, thus chemoselectivity can be achieved. Furthermore, there is no over oxidation of aldehydes to carboxylic acids as opposed to another oxidation methods such the Jones oxidation.

Potassium Permanganate

Potassium permanganate ($KMnO_4$) is a very strong oxidant able to react with many functional groups, such as secondary alcohols, 1,2-diols, aldehydes, alkenes, oximes, sulfides and thiols. Under controlled conditions, $KMnO_4$ oxidizes very efficiently primary alcohols to carboxylic acids. This reaction, which was first described in detail by Fournier, is typically carried out by adding $KMnO_4$ to a solution or suspension of the alcohol in an alkaline aqueous solution.

The resulting mixture is stirred till the oxidation is complete. For the reaction to proceed efficiently, the alcohol must be at least partially dissolved in the aqueous solution. This can be facilitated by the addition of an organic co-solvent such as dioxane, pyridine, acetone or *t*-BuOH. $KMnO_4$ will readily react with a carbon - carbon double bond before oxidizing a primary alcohol.

$KMnO_4$ is decomposed in water, resulting in formation of manganese dioxide (MnO_2) and gaseous oxygen. This decomposition is catalysed by acid, base and MnO_2. As the extent of this decomposition is difficult to estimate during the oxidation of primary alcohols, the quantity of $KMnO_4$ must be adjusted during the oxidation by adding it sequentially until the oxidation is complete.

Potassium permanganate is an inorganic chemical compound with the formula $KMnO_4$. It is a salt consisting of K^+ and MnO_4^- ions. Formerly known as permanganate of potash or Condy's crystals, it is a strong oxidizing agent. It dissolves in water to give intensely purple solutions, the evaporation of which leaves prismatic purplish-black glistening crystals. In 2000, worldwide production was estimated at 30,000 tonnes. In this compound, manganese is in the +7 oxidation state.

Structure and Preparation

Potassium permanganate is produced industrially from manganese dioxide, which also occurs as the mineral pyrolusite. The MnO_2 is fused with potassium hydroxide and heated in air or with a source of oxygen, like potassium nitrate or chlorate. This process gives potassium manganate, which upon electrolytic oxidation in alkaline media, or by boiling the manganate solution in the presence of carbon dioxide until all the green colour is discharged, gives potassium permanganate.

$$2\ MnO_2 + 4\ KOH + O_2 \rightarrow 2\ K_2MnO_4 + 2\ H_2O$$

$$2\ MnO_4^{2} + Cl_2 \rightarrow 2\ MnO_4^- + 2\ Cl^-$$

or:

$$3\ K_2MnO_4 + 2\ CO_2 \rightarrow 2\ KMnO_4 + 2\ K_2CO_3 + MnO_2$$

In which the potassium permanganate is separated by filtering the insoluble manganese dioxide, evaporating the solution to 1/3 and recrystallizing it. Permanganate salts can also be generated by treating a solution of Mn^{2+} ions with strong oxidants such as lead dioxide (PbO_2), or sodium bismuthate ($NaBiO_3$). Tests for the presence of manganese exploit the vivid violet colour of permanganate produced by these reagents. $KMnO_4$ forms orthorhombic crystals with constants: a = 910.5 pm, b = 572.0 pm, c = 742.5 pm. The overall motif is similar to that for barium sulfate, with which it forms solid solutions. In the solid (as in solution), each MnO_4^- centres are tetrahedral. The Mn-O distances are 1.62 Å.

Analytical

Potassium permanganate can be used to quantitatively determine the total oxidisable organic material in an aqueous sample. The value determined is known as the *permanganate value.* In analytical chemistry, a standardized aqueous solution of $KMnO_4$ is sometimes used as an oxidizing titrant for redox titrations (permanganometry). In a related way, it is used as a reagent to determine the Kappa number of wood pulp. For the standardization of $KMnO_4$ solutions, reduction by oxalic acid is often used.

Aqueous, acidic solutions of $KMnO_4$ are used to collect gaseous mercury in flue gas during stationary source emissions testing.

Dilute solutions of $KMnO_4$ convert alkenes into diols (glycols). This behaviour is also used as a qualitative test for the presence of double or triple bonds in a molecule, since the reaction decolourizes the permanganate solution. It is sometimes referred to as Baeyer's reagent. However, bromine serves better in measuring unsaturation (double or triple bonds) quantitatively, since $KMnO_4$, being a very strong oxidizing agent, can react with a variety of groups.

Under acidic conditions, the alkene double bond is cleaved to give the appropriate carboxylic acid:

$CH_3(CH_2)_{17}CH{=}CH_2 + 2\ KMnO_4 + 3\ H_2SO_4 \rightarrow CH_3(CH_2)_{17}COOH + CO_2 + 4\ H_2O + K_2SO_4 + 2\ MnSO_4$

Potassium permanganate oxidizes aldehydes to carboxylic acids, such as the conversion of *n*-heptanal to heptanoic acid:

$5\ C_6H_{13}CHO + 2\ KMnO_4 + 3\ H_2SO_4 \rightarrow 5\ C_6H_{13}COOH + 3\ H_2O + K_2SO_4 + 2\ MnSO_4$

Even an alkyl group (with a benzylic hydrogen) on an aromatic ring is oxidized, e.g. toluene to benzoic acid.

$5\ C_6H_5CH_3 + 6\ KMnO_4 + 9\ H_2SO_4 \rightarrow 5\ C_6H_5COOH + 14\ H_2O + 2\ K_2SO_4 + 6\ MnSO_4$

Glycols and polyols are highly reactive toward $KMnO_4$. For example, addition of potassium permanganate to an aqueous solution of sugar and sodium hydroxide produces the "chemical chameleon" reaction, which involves dramatic colour changes associated with the various oxidation states of manganese. A related vigorous reaction is exploited as a fire starter in survival kits. For example, a mixture of potassium permanganate and glycerol or pulverized glucose ignites readily. Its sterilizing properties are another reason for inclusion of $KMnO_4$ in a survival kit.

"Purple Benzene"

Purple benzene refers to solutions prepared by treating a two phase mixture of aqueous potassium permanganate and benzene with a quat salt. The quat cation forms a salt of permanganate that is soluble in benzene, giving this organic solvent a purple colour. The reaction illustrate the ability of quat salts to confer lipophilicity to hydrophilic anions. The same effect can be obtained by complexing the potassium to 18-crown-6.

Reaction with Acids

Concentrated sulfuric acid reacts with $KMnO_4$ to give Mn_2O_7, which can be explosive. Similarly concentrated hydrochloric acid gives chlorine. The Mn-containing products from redox reactions depend on the pH. Acidic solutions of permanganate are reduced to the faintly pink manganese(II) ion (Mn^{2+}) and water. In neutral solution, permanganate is only reduced by $3e^-$ to give MnO_2, wherein Mn is in a +4 oxidation state.

This is the material that stains one's skin when handling $KMnO_4$. $KMnO_4$ spontaneously reduces in an alkaline solution to green K_2MnO_4, wherein manganese is in the +6 oxidation state. A curious reaction occurs upon addition of concentrated sulfuric acid to potassium permanganate. Although no reaction may be apparent, the vapor over the mixture will ignite paper impregnated with alcohol. Potassium permanganate and sulfuric acid react to produce some ozone, which has a high oxidising power and rapidly oxidises the alcohol, causing it to combust. As the reaction also produces explosive Mn_2O_7, this should only be attempted with great care.

Handling

As an oxidizer that generates the dark brown product MnO_2, potassium permanganate rapidly stains virtually any organic material such as skin, paper, and clothing. This staining effect is used to "develop" thin layer chromatography plates. The redox reaction is used for artistic purposes as an agent to prepare paper for fast bleaching.

Lemon juice is enough to quickly remove colour from the paper and applied with a paint brush this can create interesting aesthetics. Even glassware containing solutions of $KMnO_4$ can become brown. MnO_2 can be removed by scrubbing with dilute acids (a lemon rubbed into the skin works especially well), with sodium thiosulfate or with a weak solution of hydrogen peroxide.

Uses

Almost all applications of potassium permanganate exploit its oxidizing properties. As a strong oxidant that does not generate toxic byproducts, $KMnO_4$ has many niche uses. Potassium permanganate is one of the principal chemicals used in the film and television industries to "age" props and set dressings. Its oxidising effects create "hundred year old" or "ancient" looks on hessian cloth, ropes, timber and glass. It was used on props and sets in films such as "Troy", "300" and "Indiana Jones".

Disinfectant and Water Treatment

As an oxidant, potassium permanganate can act as an antiseptic. For example, dilute solutions are used to treat canker sores (ulcers), disinfectant for the hands and treatment for mild pompholyx, dermatitis, and fungal infections of the hands or feet. Potassium permanganate is used extensively in the water treatment industry. It is used as a regeneration chemical to remove iron and hydrogen sulfide (rotten egg smell) from well water via a "Manganese Greensand" Filter. "Pot-Perm" is also obtainable at pool supply stores, is used additionally to treat waste water. Historically it was used to disinfect drinking water. It currently finds application in the control of nuisance organisms such as Zebra mussels in fresh water collection and treatment systems.

Biomedical Uses

Related to the use of $KMnO_4$ for water treatment, this salt is often employed as a specialised disinfectant for treating human and animal ailments. Permanganate washes were once used to treat gonorrhea and are still used to treat candidiasis. It can also be used to inactivate the poison strychnine.

Diagnostic Pathology

In histology, it is used to bleach melanin which obscures tissue detail.

Pre-treated with potassium permanganate, to obliterate Congo red reactivity, *was* thought to be definitive for AA amyloidosis; this is now generally considered to be unreliable.

Organic Synthesis

Aside from its use in water treatment, the other major application of $KMnO_4$ is as a reagent for the synthesis of organic compounds. Significant amounts are required for the synthesis of ascorbic acid, chloramphenicol, saccharin, isonicotinic acid, and pyrazinoic acid.

Fruit Preservation

Ethylene absorbents extend storage time of banana even at high temperatures. This effect can be exploited by packing banana in a polyethylene with potassium permanganate as an ethylene absorbent, that doubles banana lifespan up to 3–4 weeks without the need for refrigeration.

Survival Kits

Potassium permanganate is typically included in survival kits: as a fire starter, water sterilizer and for creating distress signals on snow.

In 1659, Johann Rudolf Glauber fused a mixture of the mineral pyrolusite and potassium carbonate to obtain a material that, when dissolved in water, gave a green solution (potassium manganate) which slowly shifted to violet and then finally red. This report represents the first description of the production of potassium permanganate.

Just under two hundred years later London chemist Henry Bollmann Condy had an interest in disinfectants, and marketed several products including *ozonised water*. He found that fusing pyrolusite with NaOH and dissolving it in water produced a solution with disinfectant properties.

He patented this solution, and marketed it as *Condy's Fluid*. Although effective, the solution was not very stable. This was overcome by using KOH rather than NaOH.

This was more stable, and had the advantage of easy conversion to the equally effective potassium permanganate crystals. This crystalline material was known as *Condy's crystals* or *Condy's powder*. Potassium permanganate was comparatively easy to manufacture so Condy was subsequently forced to spend considerable time in litigation in order to stop competitors from marketing products similar to *Condy's Fluid* or *Condy's Crystals*.

Early photographers used it as a component of flash powder. It is now replaced with other oxidizers, due to the instability of permanganate mixtures. Aqueous solutions of $KMnO_4$ have been used together with T-Stoff (i.e. 80% hydrogen peroxide) as propellant for the rocket plane Messerschmitt Me 163. In this application, it was known as Z-Stoff. This combination of propellants is sometimes still used in torpedoes.

Illustrative Reactions and Applications

Although useful reagent for some applications, due to the carcinogenic nature of chromium(VI), the Jones oxidation has slowly been replaced by other oxidation methods. It remains useful in organic synthesis. A variety of spectroscopic techniques, including IR can be used to monitor the progress of a Jones Oxidation reaction and confirm the presence of the oxidized product. At one time the Jones oxidation was used in primitive breathalyzers. Aminoindans, which are of pharmalogical interest, are prepared by the oxidation of the alcohol to ketone which is converted into an amino group. The alcohol is oxidized to the ketone with the Jones reagent. The reagent was once used to prepare salicylic acid, a precursor to aspirin. Methcathinone is a psychoactive stimulant that is sometimes used as an addictive recreational drug. It can be oxidized from certain alcohols using the Jones reagent.

Related Reagents

Several chromium oxides are used for related oxidations. These include Collins reagent and pyridinium chlorochromate The Sarett oxidation is a similar process.

PDC in DMF (Corey and Schmidt)

Pyridinium dichromate (PDC) is a bright-orange solid with the formulae $(C_5H_5NH)_2Cr_2O_7$ that is very often used for the oxidation of primary and secondary alcohols to aldehydes and ketones respectively. On the other hand, in 1979, Corey and Schmidt reported that reaction of *saturated* primary alcohols with PDC, *using dimethylformamide* (Me_2NCHO, DMF) as solvent, results in oxidation to carboxylic acids rather than aldehydes. Interestingly, no oxidation to carboxylic acids occurs on allylic and benzylic primary alcohols. The procedure of Corey and Schmidt for the oxidation of saturated primary alcohols to carboxylic acids is run under essentially neutral conditions.

The Cornforth reagent or pyridinium dichromate (PDC) is a pyridinium salt of dichromate with the chemical formula $[C_5H_5NH]_2[Cr_2O_7]$. This compound is named after the Australian-British chemist Sir John Warcup Cornforth (b. 1917) who introduced it in 1962. The Cornforth reagent is a strong oxidizing agent which can convert primary and secondary alcohols to aldehydes and ketones respectively. In its chemical structure and functions it is closely related to other compounds made from hexavalent chromium oxide, such as pyridinium chlorochromate and Collins reagent. Because of their toxicity, these reagents are rarely used nowadays.

Synthesis and Properties

The Cornforth reagent is prepared by slow addition of a concentrated aqueous solution of chromium trioxide to pyridine. The reaction may cause explosion, which is avoided by thoroughly dissolving the trioxide in water and cooling the solution by ice. The product is filtered, washed with acetone and dried, yielding an orange powder. The powder is stable in air, not particularly hygroscopic and has an almost neutral pH that facilitates its handling; it is only slightly acidic owing to the presence of pyridinium cations. The Cornforth reagent is readily soluble in water, dimethylformamide and dimethyl sulfoxide (DMSO). It is poorly soluble in acetone and chlorinated organic solvents, such as dichloromethane, and forms suspensions.

The Cornforth reagent is a strong oxidizing agent which can convert primary and secondary alcohols to ketones, both as a solution or suspension. This application was first mentioned in 1969, but fully developed only in 1979 by Corey and Schmidt. They mentioned that reaction of *saturated* primary alcohols with PDC, using dimethylformamide as solvent, results in oxidation to carboxylic acids rather than aldehydes. However, no oxidation to carboxylic acids occurs on allylic and benzylic primary alcohols. The oxidation is usually carried out at ambient conditions, in nearly neutral pH conditions, in dimethylformamide or dichloromethane or their mixture. The choice of solvent or their ratio affects the reaction rate; in particular, higher content of dimethylformamide results in stronger oxidation. The slow oxidation rate for some alcohols can be accelerated by the addition of molecular sieves, organic acids or acetic anhydride or of their combinations. The acceleration by molecular sieves works best when their pore diametre is about 0.3 nm, and it is apparently unrelated to their water absorption capability.

Among organic acids, acetic acid, pyridinium trifluoroacetate or pyridinium tosylate can be added, the first one being most efficient and easiest to remove. The achieved acceleration is remarkable, but the reaction inevitably turns from neutral (pH) to acidic. Comparable acceleration is achieved with acetic anhydride, which is used in sugar and nucleoside chemistry. Reaction acceleration depends not only on the additives but also on their form, so all reagents are preferred dry and freshly prepared, and PDC and molecular sieves should be finely ground. The disadvantage of the accelerators is that they may simultaneously promote several oxidation routes thereby reducing the selectivity of the reaction. In its chemical structure and functions, the Cornforth reagent is closely related to other pyridinium salts of hexavalent chromium oxide, such as pyridinium chlorochromate [PyH][CrO_3Cl] and to pyridine complexes such as the Collins reagent, $CrO_3 \cdot 2Py$ in dichloromethane and the Sarret reagent, $CrO_3 \cdot 2Py$ in pyridine.

Safety Issues

The Cornforth reagent is very toxic to aquatic life and may cause long-term damage to the environment if released in large amounts. It irritates skin and mucous membranes and may induce allergic reactions; it is carcinogenic. The maximum allowable concentration varies between 0.01 and 0.1 $mg \cdot m^{-3}$ in air depending on the country. As a strong oxidant, pyridinium dichromate promotes fires, releasing carbon monoxide, carbon dioxide and toxic metal smoke. The fire can be extinguished by water or CO_2.

TEMPO-Mediated Oxidations

(2,2,6,6-Tetramethylpiperidin-1-yl)oxyl, or (2,2,6,6-tetramethylpiperidin-1-yl)oxidanyl or TEMPO is a chemical compound with the formula $(CH_2)_3(CMe_2)_2NO$. This heterocycle is a red-orange, sublimable solid. As a stable radical, it has applications throughout chemistry and biochemistry. TEMPO was discovered by Lebedev and Kazarnowskii in 1960. It is prepared by oxidation of 2,2,6,6-tetramethylpiperidine. TEMPO is widely used as a radical trap, as a structural probe for biological systems in conjunction with electron spin resonance spectroscopy, as a reagent in organic synthesis, and as a mediator in controlled free radical polymerization. The stability of this radical is attributed to the steric protection provided by the four methyl groups adjacent to the nitroxyl group.

Application in Organic Synthesis

TEMPO is employed in organic synthesis as a catalyst for the oxidation of primary alcohols to aldehydes. The actual oxidant is the N-oxoammonium salt. In a catalytic cycle with sodium hypochlorite as the stoichiometric oxidant, hypochlorous acid generates the N-oxoammonium salt from TEMPO.

One typical reaction example is the oxidation of (S)-(-)-2-methyl-1-butanol to (S)-(+)-2-methylbutanal. 4-Methoxyphenethyl alcohol is oxidized to the corresponding carboxylic acid in a system of catalytic TEMPO and sodium hypochloritc and a stoichiometric amount of sodium chlorite. TEMPO oxidations also exhibit chemoselectivity, being inert towards secondary alcohols, but the reagent will convert aldehydes to carboxylic acids.

In cases where secondary oxidizing agents cause side reactions, it is possible to stoichiometrically convert TEMPO to the oxoammonium salt in a separate step. For example, in the oxidation of geraniol to geranial, 4-acetamido-TEMPO is first oxidized to the oxoammonium tetrafluoroborate.

Industrial Applications

TEMPO itself is relatively inexpensive, but there are TEMPO derivatives that are often used such as 4-hydroxy-TEMPO (TEMPOL) or 4-acetamido-TEMPO that have cheaper precursors. Examples of TEMPO use in chemical industry are bisnoralcohol (a steroid) to bisnoraldehyde conversion by Upjohn and retinol to retinal conversion by Novartis. One industrial method employs $H_5PV_2Mo_{10}O_{40}$ as co-oxidant, the reduced form of which can be reoxidized by atmospheric oxygen. Polymer-supported TEMPO catalysts are also commercially available.

Two-Step Oxidation of Alcohols to Acids via Isolated Aldehydes

As a lot of the aforementioned conditions for the oxidations of primary alcohols to acids are harsh and not compatible with common protection groups, organic chemists often use a two-step procedure for the oxidation to acids. The primary alcohol is oxidized to an aldehyde using one of the many existing procedures (e.g. IBX oxidation, Dess-Martin periodinane). The aldehyde can then be subjected to the conditions of the Pinnick oxidation using sodium chlorite. This sequence is often used in natural product synthesis, Nicolaou et al. applied it in their synthesis of Platencin.

Toxicity

Ethanol in alcoholic beverages has been consumed by humans since prehistoric times for a variety of hygienic, dietary, medicinal, religious, and recreational reasons. The consumption of large doses of ethanol causes drunkenness (intoxication), which may lead to a hangover as its effects wear off. Depending upon the dose and the regularity of its consumption, ethanol can cause acute respiratory failure or death. Because ethanol impairs judgment in humans, it can be a catalyst for reckless or irresponsible behaviour. The LD_{50} of ethanol in rats is 10.3 g/kg. Ethanol's toxicity is largely caused by its primary metabolite; acetaldehyde and secondary metabolite; acetic acid. All primary alcohols are broken down into aldehydes then to carboxylic acids, and whose toxicities are similar to acetaldehyde and acetic acid. Metabolite toxicity is reduced in rats fed N-acetylcysteine and thiamine.

Some secondary and tertiary alcohols are less poisonous than ethanol because the liver is unable to metabolise them into these toxic by-products. This makes them more suitable for recreational and medicinal use as the chronic harms are lower. Ethchlorvynol is a good example of a tertiary alcohol which saw both medicinal and recreational use.

Other alcohols are substantially more poisonous than ethanol, partly because they take much longer to be metabolized and partly because their metabolism produces substances that are even more toxic.

Methanol (wood alcohol), for instance, is oxidized to formaldehyde and then to the poisonous formic acid in the liver by alcohol dehydrogenase and formaldehyde dehydrogenase enzymes, respectively; accumulation of formic acid can lead to blindness or death. Likewise, poisoning due to other alcohols such as ethylene glycol or diethylene glycol are due to their metabolites, which are also produced by alcohol dehydrogenase. An effective treatment to prevent toxicity after methanol or ethylene glycol ingestion is to administer ethanol. Alcohol dehydrogenase has a higher affinity for ethanol, thus preventing methanol from binding and acting as a substrate. Any remaining methanol will then have time to be excreted through the kidneys.

Methanol itself, while poisonous, has a much weaker sedative effect than ethanol. Some longer-chain alcohols such as n-propanol, isopropanol, n-butanol and t-butanol do, however, have stronger sedative effects, but also have higher toxicity than ethanol.

These longer chain alcohols are found as contaminants in some alcoholic beverages and are known as fusel alcohols, and are reputed to cause severe hangovers although it is unclear if the fusel alcohols are actually responsible. Many longer chain alcohols are used in industry as solvents and are occasionally abused by alcoholics, leading to a range of adverse health effects.

Short-term effects of alcohol on the human body can take many forms. The drug alcohol, to be specific ethanol, is a central nervous system depressant with a range of side-effects. Cell membranes are highly permeable to alcohol, so once alcohol is in the bloodstream it can diffuse into nearly every biological tissue of the body. The concentration of alcohol in blood is usually measured in terms of the blood alcohol content. The amount and circumstances of consumption play a large part in determining the extent of intoxication; for example, consuming alcohol after a heavy meal causes alcohol to absorb more slowly. Hydration also plays a role, especially in determining the extent of hangovers.

After excessive drinking, unconsciousness can occur and extreme levels of consumption can lead to alcohol poisoning and death (a concentration in the blood stream of 0.40% will kill half of those affected). Alcohol may also cause death by asphyxiation from vomit.

Alcohol is an addictive drug that can greatly exacerbate sleep problems. During abstinence, residual disruptions in sleep regularity and sleep patterns are the greatest predictors of relapse.

Moderate alcohol consumption 30–60 minutes before sleep, although decreasing sleep onset latency, disrupts sleep architecture. Rebound effects occur once the alcohol has been largely metabolized, causing late night disruptions in sleep maintenance. Under conditions of moderate alcohol consumption where blood alcohol levels average 0.06–0.08 percent and decrease 0.01–0.02 percent per hour, an alcohol clearance rate of 4–5 hours would coincide with disruptions in sleep maintenance in the second half of an 8 hour sleep episode.

In terms of sleep architecture, moderate doses of alcohol facilitate "rebounds" in rapid eye movement (REM) and stage 1 sleep; following suppression in REM and stage 1 sleep in the first half of an 8 hour sleep episode, REM and stage 1 sleep increase well beyond baseline in the second half.

Moderate doses of alcohol also very quickly increase slow wave sleep (SWS) in the first half of an 8 hour sleep episode. Enhancements in REM sleep and SWS following moderate alcohol consumption are mediated by reductions in glutamatergic activity by adenosine in the central nervous system. In addition, tolerance to changes in sleep maintenance and sleep architecture develops within 3 days of alcohol consumption before bedtime.

Alcohol Consumption and Sleep Improvements

Low doses of alcohol (one 360 ml (13 imp fl oz; 12 US fl oz) beer) appear to increase total sleep time and reduce awakening during the night. The sleep-promoting benefits of alcohol dissipate at moderate and higher doses of alcohol.

Previous experience with alcohol also influences the extent to which alcohol positively or negatively affects sleep.

Under free-choice conditions, in which subjects chose between drinking alcohol or water, inexperienced drinkers were sedated while experienced drinkers were stimulated following alcohol consumption. In insomniacs, moderate doses of alcohol improve sleep maintenance.

Alcohol Consumption and Fatigue

Conditions of sleep deprivation correlate positively with increased alcohol consumption. In Northern climates, increased alcohol consumption during the winter is attributed to escalations in fatigue.

Alcohol Abstinence and Sleep Disruptions

Hormonal imbalance and sleep disruption following withdrawal from chronic alcohol consumption are strong predictors of relapse. During abstinence, recovering alcoholics have attenuated melatonin secretion at onset of a sleep episode, resulting in prolonged sleep onset latencies. Escalations in cortisol and core body temperatures during the sleep period contribute to poor sleep maintenance.

Alcohol Consumption and Balance

Alcohol affects balance by changing the viscosity of the endolymph within the otolithic membrane, the fluid inside the semicircular canals inside the ear. The endolymph surrounds the cupula which contains hair cells within the semicircular canals.

When the head is tilted, the endolymph flows and moves the cupula. The hair cells then bend and send signals to the brain indicating the direction in which the head is tilted. By changing the viscosity of the endolymph to become less dense when alcohol enters the system, the hair cells can move more easily within the ear, which sends the signal to the brain and results in exaggerated and overcompensated movements of body. This can also result in vertigo, or "the spins."

Moderate Doses

Ethanol inhibits the ability of glutamate to open the cation channel associated with the N-methyl-D-aspartate (NMDA) subtype of glutamate receptors. Stimulated areas include the cortex, hippocampus and nucleus accumbens, which are responsible for thinking and pleasure seeking.

Another one of alcohol's agreeable effects is body relaxation, possibly caused by neurons transmitting electrical signals in an alpha waves-pattern; such waves are observed (with the aid of EEGs) when the body is relaxed.

Short-term effects of alcohol include the risk of injuries, violence and fetal damage. Alcohol has also been linked with lowered inhibitions, though it is unclear to what degree this is chemical versus psychological as studies with placebos can often duplicate the social effects of alcohol at low to moderate doses. Some studies have suggested that intoxicated people have much greater control over their behaviour than is generally recognised, though they have a reduced ability to evaluate the consequences of their behaviour. Behavioural changes associated with drunkenness are, to some degree, contextual. A scientific study found that people drinking in a social setting significantly and dramatically altered their behaviour immediately after the first sip of alcohol, well before the chemical itself could have filtered through to the nervous system.

Areas of the brain responsible for planning and motor learning are dulled. A related effect, caused by even low levels of alcohol, is the tendency for people to become more animated in speech and movement.

This is due to increased metabolism in areas of the brain associated with movement, such as the nigrostriatal pathway. This causes reward systems in the brain to become more active, which may induce certain individuals to behave in an uncharacteristically loud and cheerful manner.

Alcohol has been known to mitigate the production of antidiuretic hormone, which is a hormone that acts on the kidney to favour water reabsorption in the kidneys during filtration. This occurs because alcohol confuses osmoreceptors in the hypothalamus, which relay osmotic pressure information to the posterior pituitary, the site of antidiuretic hormone release. Alcohol causes the osmoreceptors to signal that there is low osmotic pressure in the blood, which triggers an inhibition of the antidiuretic hormone. As a consequence, one's kidneys are no longer able to reabsorb as much water as they should be absorbing, leading to creation of excessive volumes of urine and the subsequent overall dehydration.

Excessive Doses

Acute alcohol intoxication through excessive doses in general causes short- or long-term health effects. NMDA receptors start to become unresponsive, slowing areas of the brain for which they are responsible. Contributing to this effect is the activity that alcohol induces in the gamma-aminobutyric acid (GABA) system. The GABA system is known to inhibit activity in the brain. GABA could also be responsible for the memory impairment that many people experience. It has been asserted that GABA signals interfere with the *registration* and *consolidation* stages of memory formation. As the GABA system is found in the hippocampus (among other areas in the CNS), which is thought to play a large role in memory formation, this is thought to be possible. Anterograde amnesia, colloquially referred to as "blacking out", is another symptom of heavy drinking. This is the loss of memory during and after an episode of drinking. When alcohol is consumed at a rapid rate, the point at which most healthy people's long-term memory creation starts to fail usually occurs at approximately 0.20% BAC, but can be reached as low as 0.14% BAC for inexperienced drinkers.

Another classic finding of alcohol intoxication is ataxia, in its appendicular, gait, and truncal forms. Appendicular ataxia results in jerky, uncoordinated movements of the limbs, as though each muscle were working independently from the others. Truncal ataxia results in postural instability; gait instability is manifested as a disorderly, wide-based gait with inconsistent foot positioning. Ataxia is responsible for the observation that drunk people are clumsy, sway back and forth, and often fall down. It is presumed to be due to alcohol's effect on the cerebellum.

Pathophysiology

At low or moderate doses, alcohol acts primarily as a positive allosteric modulator of $GABA_A$. Alcohol binds to several different subtypes of $GABA_A$, but not to others. The main subtypes responsible for the subjective effects of alcohol are the $\alpha1\beta3\gamma2$, $\alpha5\beta3\gamma2$, $\alpha4\beta3\delta$ and $\alpha6\beta3\delta$ subtypes, although other subtypes such as $\alpha2\beta3\gamma2$ and $\alpha3\beta3\gamma2$ are also affected.

Activation of these receptors causes most of the effects of alcohol such as relaxation and relief from anxiety, sedation, ataxia and increase in appetite and lowering of inhibitions that can cause a tendency toward violence in some people.

Alcohol has a powerful effect on glutamate as well. Alcohol decreases glutamate's ability to bind with NMDA and acts as an antagonist of the NMDA receptor, which plays a critical role in LTP by allowing Ca2+ to enter the cell. These inhibitory effects are thought to be responsible for the "memory blanks" that can occur at levels as low as 0.03% blood level.

In addition, reduced glutamate release in the dorsal hippocampus has been linked to spatial memory loss. Chronic alcohol users experience an upregulation of NMDA receptors because the brain is attempting to reestablish homeostasis. When a chronic alcohol user stops drinking for more than 10 hours, apoptosis can occur due to excitotoxicity. The seizures experienced during alcohol abstinence are thought to be a result of this NMDA upregulation. Alteration of NMDA receptor numbers in chronic alcoholics is likely to be responsible for some of the symptoms seen in delirium tremens during severe alcohol withdrawal, such as delirium and hallucinations.

Other targets such as sodium channels can also be affected by high doses of alcohol, and alteration in the numbers of these channels in chronic alcoholics is likely to be responsible for as well as other effects such as cardiac arrhythmia. Other targets that are affected by alcohol include cannabinoid, opioid and dopamine receptors, although it is unclear whether alcohol affects these directly or if they are affected by downstream consequences of the GABA/NMDA effects. People with a family history of alcoholism may exhibit genetic differences in the response of their NMDA glutamate receptors as well as the ratios of GABA-A subtypes in their brain.

Animal models using mammals and invertebrates have been informative in studying the effects of ethanol on not only pharmacokinetics of alcohol but also pharmacodynamics, in particular in the nervous system. Ethanol-induced intoxication is not uncommon in the animal kingdom, as noted here:

"Many of us have noticed that bees or yellow jackets cannot fly well after having drunk the juice of overripe fruits or berries; bears have been seen to stagger and fall down after eating fermented honey; and birds often crash or fly haphazardly while intoxicated on ethanol that occurs naturally as free-floating microorganisms convert vegetable carbohydrates to alcohol." More recently, studies using animal models have begun to elucidate the effects of ethanol on the nervous system. Traditionally, many studies have been performed in mammals, such as mice, rats, and non-human primates.

However, non-mammalian animal models have also been employed; in particular, Ulrike Heberlein group at UC San Francisco has used *Drosophila melanogaster*, the fruit fly, taking advantage of its facile genetics to dissect the neural circuits and molecular pathways, upon which ethanol acts.

The series of studies carried in the Heberlein lab has identified insulin and its related signaling pathways as well as biogenic amines in the invertebrate nervous system as being important in alcohol tolerance. The value of antabuse (disulfiram) as a treatment for alcoholism has been tested using another invertebrate animal model, the honey bees. It is important to note that some of the analogous biochemical pathways and neural systems have been known to be important in alcohol's effects on humans, while the possibility that others may also be important remains unknown. Research of alcohol's effects on the nervous system remains a hot topic of research, as scientists inch toward understanding the problem of alcohol addiction.

In addition to the studies carried out in invertebrates, researchers have also used vertebrate animal models to study various effects of ethanol on behaviours.

The long term effects of alcohol range from possible health benefits for low levels of alcohol (ethanol) consumption to severe detrimental effects in cases of chronic alcohol abuse. There is a strong correlation between 'high levels' of alcohol consumption and an increased risk of developing alcoholism, cardiovascular disease, malabsorption, chronic pancreatitis, alcoholic liver disease, and cancer. Damage to the central nervous system and peripheral nervous system can occur from chronic alcohol abuse. Long-term use of alcohol in excessive quantities is capable of damaging nearly every organ and system in the body. The developing adolescent brain is particularly vulnerable to the toxic effects of alcohol, as is the developing brain of the unborn, possibly resulting in the fetal alcohol syndrome (FAS).

Historically doctors have promoted alcohol for its perceived health benefits and most recently for protection against coronary heart disease. This is known as the French paradox.

There is evidence of cardiovascular benefits from drinking 1 - 2 drinks per day; however, the health benefits from moderate intake of alcohol are controversial. Alcohol should be regarded as a recreational drug with potentially serious adverse effects on health and it is not recommended for cardio-protection in the place of safer and proven traditional methods such as exercise and proper nutrition.

Some experts argue that the benefits of moderate alcohol consumption may be outweighed by other increased risks, including those of injuries, violence, fetal damage, certain forms of cancer, liver disease and hypertension.

As the apparent health benefits of moderate alcohol consumption are limited for populations at low risk of heart disease, other experts urge caution because of the possibility that recommending moderate alcohol consumption may lead to an increased risk of alcohol abuse.

The adverse effects of long-term excessive use of alcohol are close to those seen with other sedative-hypnotics (apart from organ toxicity which is much more problematic with alcohol). Withdrawal effects and dependence are also almost identical.

Alcohol at moderate levels has some positive and negative effects on health. The negative effects include increased risk of liver diseases, oropharyngeal cancer, esophageal cancer and pancreatitis. Conversely moderate intake of alcohol may have some beneficial effects on gastritis and cholelithiasis. Chronic alcohol misuse and abuse has serious effects on physical and mental health.

Chronic excess alcohol intake, or alcohol dependence, can lead to a wide range of neuropsychiatric or neurological impairment, cardiovascular disease, liver disease, and malignant neoplasms.

The psychiatric disorders which are associated with alcoholism include major depression, dysthymia, mania, hypomania, panic disorder, phobias, generalised anxiety disorder, personality disorders, schizophrenia, suicide, neurologic deficits (e.g. impairments of working memory, emotions, executive functions, visuospatial abilities and gait and balance) and brain damage. Alcohol dependence is associated with hypertension, coronary heart disease, and ischemic stroke, cancer of the respiratory system, and also cancers of the digestive system, liver, breast and ovaries. Heavy drinking is associated with liver disease, such as cirrhosis. Studies have focused on both men and women, various age groups, and people of many ethnic groups.

Published papers now total in the many hundreds, with studies having shown correlation between moderate alcohol use and health that may instead have been due to the beneficial effects of socialization that is often accompanied by alcohol consumption. Some of the specific ways alcohol affects cardiovascular health have been studied. The impact of alcohol on aging is negative when excessive amounts of alcohol are consumed.

Maximum Quantity Recommended

Different countries recommend different maximum quantities. For most countries, the maximum quantity for men is 140 g–210 g per week. For women, the range is 84 g–140 g per week. Most countries recommend total abstinence whilst pregnant or breastfeeding.

Cardiovascular System

A meta-analysis of 34 studies found a reduced risk of mortality from coronary heart disease in men who drank 2 - 4 drinks per day and women who drank 1 - 2 drinks per day.

A meta-analysis of randomized trials found that alcohol consumption in moderation decreases serum levels of fibrinogen, a protein that promotes clot formation and increases levels of tissue type plasminogen activator, an enzyme that helps dissolve clots. The serum levels of C-reactive protein (CRP), a marker of inflammation and predictor of CHD (coronary heart disease) risk, are lower in people who drink moderately than those who abstain from alcohol suggesting that alcohol consumption in moderation might have anti-inflammatory effects.

In addition to its psychotropic properties, alcohol has anticoagulation properties similar to warfarin. Additionally, thrombosis is lower among moderate drinkers than teetotalers.

Despite epidemiological evidence, some criticize the idea of recommending alcohol for health benefits. A doctor at the World Health Organisation stated that recommending moderate alcohol consumption for health benefits is “ridiculous and dangerous”.

There have been no randomised controlled trials that have demonstrated cardiac benefits of alcohol. Due to the risks of abuse, dependence, and adverse effects, alcohol should never be recommended for cardiac benefits as a substitute to well-proven measures, such as a good diet, exercise or pharmaceutical drugs. It has been argued that the health benefits from alcohol are at best debatable and may have been exaggerated by the alcohol industry. Alcohol should be regarded as a recreational drug with potentially serious adverse effects on health and should not be promoted for cardio-protection.

Peripheral Arterial Disease

"Moderate alcohol consumption appears to decrease the risk of PAD in apparently healthy men." "In this large population-based study, moderatable alcohol consumption was inversely associated with peripheral arterial disease in women but not in men. Residual confounding by smoking may have influenced the results.

Among nonsmokers an inverse association was found between alcohol consumption and peripheral arterial disease in both men and women."

Intermittent Claudication (IC)

A study found that moderate consumption of alcohol had a protective effect against intermittent claudication. The lowest risk was seen in men who drank 1 to 2 drinks per day and in women who drank half to 1 drink per day.

Heart Attack and Stroke

Drinking in moderation has been found to help those who have suffered a heart attack survive it. However, excessive alcohol consumption leads to an increased risk of heart failure. A review of the literature found that half a drink of alcohol offered the best level of protection.

However, they noted that at present there have been no randomised trials to confirm the evidence which suggests a protective role of low doses of alcohol against heart attacks. However, moderate alcohol consumption is associated with hypertension. There is an increased risk of hypertriglyceridemia, cardiomyopathy, hypertension, and stroke if 3 or more standard drinks of alcohol are taken per day.

Cardiomyopathy

Large amount of alcohol over the long term can lead to alcoholic cardiomyopathy. Alcoholic cardiomyopathy presents in a manner clinically identical to idiopathic dilated cardiomyopathy, involving hypertrophy of the musculature of the heart that can lead to congestive heart failure.

Hematologic Diseases

Alcoholics may have anemia from several causes; they may also develop thrombocytopenia from direct toxic effect on megakaryocytes, or from hypersplenism.

Nervous System

Chronic heavy alcohol consumption impairs brain development, causes brain shrinkage, dementia, physical dependence, increases neuropsychiatric and cognitive disorders and causes distortion of the brain chemistry.

Some studies however have shown that moderate alcohol consumption may decrease risk of dementia, including Alzheimer disease, although there are studies which find the opposite. At present, due to poor study design, and methodology the literature is inconclusive on whether moderate alcohol consumptions increases the risk of dementia or decreases it.

Evidence for a protective effect of low to moderate alcohol consumption on age related cognitive decline and dementia has been suggested by some research, however, other research has not found a protective effect of low to moderate alcohol consumption. Some evidence suggests that low to moderate alcohol consumption may speed up brain volume loss. Chronic consumption of alcohol may result in increased plasma levels of the toxic amino acid homocysteine; which may explain alcohol withdrawal seizures, alcohol-induced brain atrophy and alcohol-related cognitive disturbances. Alcohol's impact on the nervous system can also include disruptions of memory and learning, such as resulting in a blackout phenomenon.

Strokes

Epidemiological studies of middle-aged populations generally find the relationship between alcohol intake and the risk of stroke to be either U- or J-shaped.

"The consistency in the vascular benefit associated with moderate drinking (compared with non-drinking) observed across different studies, together with the existence of credible biological pathways, strongly suggests that at least some of this benefit is real. However, because of biases introduced by: choice of reference categories; reverse causality bias; variations in alcohol intake over time; and confounding, some of it is likely to be an artefact." Drinking lots of alcohol has been negatively linked to stroke.

According to the National Stroke Association, more than 2 drinks per day (one drink is ~5 ounces of wine) may increase stroke risk by 50% and lead to other medical problems. Because of this and other known harmful effects of heavy drinking, care should be taken when considering safe amounts of alcohol intake.

Alcohol abuse is associated with wide spread and significant brain lesions. Alcohol related brain damage is not only due to the direct toxic effects of alcohol; alcohol withdrawal, nutritional deficiency, electrolyte disturbances, and liver damage are also believed to contribute to alcohol related brain damage.

Adolescent Brain Development

Consuming large amounts of alcohol over a period of time can impair normal brain development in humans. Deficits in retrieval of verbal and nonverbal information and in visuospatial functioning were evident in youths with histories of heavy drinking during early and middle adolescence.

During adolescence critical stages of neurodevelopment occur, including remodelling and functional changes in synaptic plasticity and neuronal connectivity in different brain regions.

These changes may make adolescents especially susceptible to the harmful effects of alcohol. Compared to adults, adolescents exposed to alcohol are more likely to exhibit cognitive deficits (including learning and memory dysfunction). Some of these cognitive effects, such as learning impairments, may persist into adulthood.

Mechanisms of Action

Neuroinflammation

Ethanol can trigger the activation of astroglial cells which can produce a proinflammatory response in the brain. Ethanol interacts with the TLR4 and IL-1RI receptors on these cells to activate intracellular signal transduction pathways. Specifically, ethanol induces the phosphorylation of IL-1R-associated kinase (IRAK), ERK1/2, stress-activated protein kinase (SAPK)/JNK, and p38 mitogen-activated protein kinase (p38 MAPK).

Activation of the IRAK/MAPK pathway leads to the stimulation of the transcription factors NF-kappaB and AP-1. These transcription factors cause the upregulation of inducible nitric oxide synthase (iNOS) and cyclooxygenase-2 (COX-2) expression.

The upregulation of these inflammatory mediators by ethanol is also associated with an increase in caspase 3 activity and a corresponding increase in cell apoptosis.

The exact mechanism by which various concentrations of ethanol either activates or inhibits TLR4/IL-1RI signaling is not currently known, though it may involve alterations in lipid raft clustering or cell adhesion complexes and actin cytoskeleton organisation.

Changes in Dopaminegric and Glutamatergic Signaling Pathways

Intermittent ethanol treatment causes a decrease in expression of the dopamine receptor type 2 (D2R) and a decrease in phosphorylation of 2B subunit of the NMDA receptor (NMDAR2B) in the prefrontal cortex, hippocampus, nucleus accumbens, and for only D2R the striatum.

It also causes changes in the acetylation of histones H3 and H4 in the prefrontal cortex, nucleus accumbens, and striatum, suggesting chromatin remodelling changes which may mediate long-term alterations.

Additionally, adolescent rats pre-exposed to ethanol have higher basal levels of dopamine in the nucleus accumbens, along with a prolonged dopamine response in this area in response to a challenge dose of ethanol. Together, these results suggest that alcohol exposure during adolescence can sensitize the mesolimbic and mesocortical dopamine pathways to cause changes in dopaminegric and glutamatergic signaling, which may affect the remodelling and functions of the adolescent brain. These changes are significant as alcohol's effect on NMDARs could contribute to learning and memory dysfunction.

Inhibition of Hippocampal Neurogenesis

Excessive alcohol intake (binge drinking) causes a decrease in hippocampal neurogenesis, via decreases in neural stem cell proliferation and newborn cell survival. Alcohol decreases the number of cells in S-phase of the cell cycle, and may arrest cells in the G1 phase, thus inhibiting their proliferation. Ethanol has different effects on different types of actively dividing hippocampal progenitors during their initial phases of neuronal development.

Chronic alcohol exposure decreases the number of proliferating cells that are radial glia-like, preneuronal, and intermediate types, while not affecting early neuronal type cells; suggesting ethanol treatment alters the precursor cell pool.

Furthermore, there is a greater decrease in differentiation and immature neurons than there is in proliferating progenitors, suggesting that the abnormal decrease in the percentage of actively dividing preneuronal progenitors results in a greater reduction in the maturation and survival of postmitotic cells. Additionally, alcohol exposure increased several markers of cell death. In these studies neural degeneration seems to be mediated by non-apoptotic pathways. One of the proposed mechanisms for alcohol's neurotoxicity is the production of nitric oxide (NO), yet other studies have found alcohol-induced NO production to lead to apoptosis.

Cognition and Dementia

Excessive alcohol intake is associated with impaired prospective memory. This impaired cognitive ability leads to increased failure to carry out an intended task at a later date, for example, forgetting to lock the door or to post a letter on time. The higher the volume of alcohol consumed and the longer consumed, the more severe the impairments. One of the organs most sensitive to the toxic effects of chronic alcohol consumption is the brain. In France approximately 20% of admissions to mental health facilities are related to alcohol related cognitive impairment, most notably alcohol related dementia. Chronic excessive alcohol intake is also associated with serious cognitive decline and a range of neuropsychiatric complications.

The elderly are the most sensitive to the toxic effects of alcohol on the brain. There is some inconclusive evidence that small amounts of alcohol taken in earlier adult life are protective in later life against cognitive decline and dementia. However, a study concluded, "Our findings suggest that, despite previous suggestions, moderate alcohol consumption does not protect older people from cognitive decline."

Acetaldehyde is produced by the liver during breakdown of ethanol. People who have a genetic deficiency for the subsequent conversion of acetaldehyde into acetic acid (a trait more prevalent in those of East Asian descent) may have a greater risk of Alzheimer's disease. "These results indicate that the ALDH2 deficiency is a risk factor for LOAD [late-onset Alzheimer's disease] ..."

Wernicke-Korsakoff syndrome is a manifestation of thiamine deficiency, usually as a secondary effect of alcohol abuse. The syndrome is a combined manifestation of two eponymous disorders, Korsakoff's Psychosis and Wernicke's encephalopathy, named after Drs. Sergei Korsakoff and Carl Wernicke. Wernicke's encephalopathy is the acute presentation of the syndrome and is characterised by a confusional state while Korsakoff's psychosis main symptoms are amnesia and executive dysfunction.

Essential Tremor

Essential tremors can be temporarily relieved in up to two-thirds of patients by drinking small amounts of alcohol. Ethanol is known to activate aminobutyric acid type A (GABAA) and inhibit N-methyl-D-aspartate (NMDA) glutamate receptors, which are both implicated in essential tremor pathology and could underlie the ameliorative effects . Additionally, the effects of ethanol have been studied in different animal essential tremor models. Chronic use of alcohol used to induce sleep can lead to insomnia. Frequent moving between sleep stages occurs, with awakenings due to headaches and diaphoresis. Stopping chronic alcohol abuse can also lead to profound disturbances of sleep with vivid dreams. Chronic alcohol abuse is associated with NREM stage 3 and 4 sleep as well as suppression of REM sleep and REM sleep fragmentation. During withdrawal REM sleep is typically exaggerated as part of a rebound effect.

High rates of major depressive disorder occur in heavy drinkers and those who abuse alcohol. Whether it is more true that major depressive disorder causes self-medicating alcohol abuse, or the increased incidence of the disorder in alcohol abusers is caused by the drinking, is not known though some evidence suggests drinking causes the disorder. Alcohol misuse is associated with a number of mental health disorders and alcoholics have a very high suicide rate. A study of people hospitalised for suicide attempts found that those who were alcoholics were 75 times more likely to go on to successfully commit suicide than non-alcoholic suicide attempters.

In the general alcoholic population the increased risk of suicide compared to the general public is 5 - 20 times greater. About 15 percent of alcoholics commit suicide. Abuse of other drugs is also associated with an increased risk of suicide. About 33 percent of suicides in the under 35s are due to alcohol or other substance misuse.

Social skills are significantly impaired in people suffering from alcoholism due to the neurotoxic effects of alcohol on the brain, especially the prefrontal cortex area of the brain. The social skills that are impaired by alcohol abuse include impairments in perceiving facial emotions, prosody perception problems and theory of mind deficits; the ability to understand humour is also impaired in alcohol abusers. Studies have shown that alcohol dependence relates directly to cravings and irritability. Another study has shown that alcohol use is a significant predisposing factor towards antisocial behaviour in children. Depression, anxiety and panic disorder are disorders commonly reported by alcohol dependent people. Alcoholism is associated with dampened activation in brain networks responsible for emotional processing (*e.g.* the amygdala and hippocampus). Evidence that the mental health disorders are often induced by alcohol misuse via distortion of brain neurochemistry is indicated by the improvement or disappearance of symptoms that occurs after prolonged abstinence, although problems may worsen in early withdrawal and recovery periods. Psychosis is secondary to several alcohol-related conditions including acute intoxication and withdrawal after significant exposure. Chronic alcohol misuse can cause psychotic type symptoms to develop, more so than with other drugs of abuse. Alcohol abuse has been shown to cause an 800% increased risk of psychotic disorders in men and a 300% increased risk of psychotic disorders in women which are not related to pre-existing psychiatric disorders. This is significantly higher than the increased risk of psychotic disorders seen from cannabis use making alcohol abuse a very significant cause of psychotic disorders. Prominent hallucinations and/or delusions are usually present when a patient is intoxicated or recently withdrawn from alcohol.

While alcohol initially helps social phobia or panic symptoms, with longer term alcohol misuse can often worsen social phobia symptoms and can cause panic disorder to develop or worsen, during alcohol intoxication and especially during the alcohol withdrawal syndrome. This effect is not unique to alcohol but can also occur with long term use of drugs which have a similar mechanism of action to alcohol such as the benzodiazepines which are sometimes prescribed as tranquillisers to people with alcohol problems.

Approximately half of patients attending mental health services for conditions including anxiety disorders such as panic disorder or social phobia suffer from alcohol or benzodiazepine dependence. It was noted that every individual has an individual sensitivity level to alcohol or sedative hypnotic drugs and what one person can tolerate without ill health another will suffer very ill health and that even moderate drinking can cause rebound anxiety syndromes and sleep disorders. A person who is suffering the toxic effects of alcohol will not benefit from other therapies or medications as they do not address the root cause of the symptoms.

Digestive System and Weight Gain

The impact of alcohol on weight-gain is contentious: some studies find no effect, others find decreased or increased effect on weight gain.

Alcohol use increases the risk of chronic gastritis (stomach inflammation); it is one cause of cirrhosis, hepatitis, and pancreatitis in both its chronic and acute forms.

Metabolic Syndrome

A study concluded, “Mild to moderate alcohol consumption is associated with a lower prevalence of the metabolic syndrome, with a favourable influence on lipids, waist circumference, and fasting insulin. This association was strongest among whites and among beer and wine drinkers.” This is also true for Asians.

A J-curve association between alcohol intake and metabolic syndrome was found: "The results of the present study suggest that the metabolic syndrome is negatively associated with light alcohol consumption (1–15 g alcohol/d) in Korean adults". However, "odds ratios for the metabolic syndrome and its components tended to increase with increasing alcohol consumption."

Gallbladder Effects

Research has found that drinking reduces the risk of developing gallstones. Compared with alcohol abstainers, the relative risk of gallstone disease, controlling for age, sex, education, smoking, and body mass index, is 0.83 for occasional and regular moderate drinkers (< 25 ml of ethanol per day), 0.67 for intermediate drinkers (25-50 ml per day), and 0.58 for heavy drinkers.

This inverse association was consistent across strata of age, sex, and body mass index." Frequency of drinking also appears to be a factor. "An increase in frequency of alcohol consumption also was related to decreased risk. Combining the reports of quantity and frequency of alcohol intake, a consumption pattern that reflected frequent intake (5-7 days/ week) of any given amount of alcohol was associated with a decreased risk, as compared with nondrinkers. In contrast, infrequent alcohol intake (1-2 days/week) showed no significant association with risk."

Consumption of alcohol is unrelated to gallbladder disease. However one study suggested that drinkers who take vitamin C (ascorbic acid) might reduce their risk of gallbladder disease.

Alcoholic liver disease is a major public health problem. For example in the United States up to two million people have alcohol related liver disorders. Chronic alcohol abuse can cause fatty liver, cirrhosis and alcoholic hepatitis. Treatment options are limited and consist of most importantly discontinuing alcohol consumption. In cases of severe liver disease, the only treatment option may be a liver transplant in alcohol abstinent patients. Research is being conducted into the effectiveness of anti-TNFs. Certain complementary medications, e.g., milk thistle and silymarin, appear to offer some benefit. Alcohol is a leading cause of liver cancer in the Western world, accounting for 32-

45% of hepatic cancers. Up to half a million people in the United States develop alcohol related liver cancer. Moderate alcohol consumption also increases the risk of liver disease.

Alcohol misuse is a leading cause of both acute pancreatitis and chronic pancreatitis. Chronic excessive intake of alcohol can cause destruction of the pancreas resulting in severe chronic pain, which may progress to pancreatic cancer. Chronic pancreatitis often results in malabsorption problems and diabetes. Chronic alcohol ingestion impairs multiple critical cellular functions in the lung.

These cellular impairments lead to increased susceptibility to serious complications from lung disease. Recent research cites alcoholic lung disease as comparable to liver disease in alcohol related mortality. Alcoholics have a higher risk of developing acute respiratory distress syndrome (ARDS) and experience higher rates of mortality from ARDS when compared to non-alcoholics.

Research indicates that drinking alcohol is associated with a lower risk of developing kidney stones. One study concludes, "Since beer seemed to be protective against kidney stones, the physiologic effects of other substances besides ethanol, especially those of hops, should also be examined." "...consumption of coffee, alcohol, and vitamin C supplements were negatively associated with stones."

"After mutually adjusting for the intake of other beverages, the risk of stone formation decreased by the following amount for each 240-ml (8-oz) serving consumed daily: caffeinated coffee, 10%; decaffeinated coffee, 10%; tea, 14%; beer, 21%; and wine, 39%." "...stone formation decreased by the following amount for each 240-mL (8-oz) serving consumed daily: 10% for caffeinated coffee, 9% for decaffeinated coffee, 8% for tea, and 59% for wine." (CI data excised from last two quotes.).

Long term excessive intake of alcohol can lead to damage to the central nervous system and the peripheral nervous system resulting in loss of sexual desire and impotence in men. This can result due to a reduction of testosterone from ethanol-induced testicular atrophy, resulting in increased feminisation of males and is a clinical feature of alcohol abusing males who have cirrhosis of the liver.

Diabetes Mellitus

Moderate drinkers may have a lower risk of diabetes than non-drinkers. On the other hand, binge drinking and high alcohol consumption may increase the risk of type 2 diabetes in women." Alcohol consumption promotes insulin sensitivity.

Rheumatoid Arthritis

Regular consumption of alcohol is associated with an increased risk of gouty arthritis and a decreased risk of rheumatoid arthritis. Two recent studies report that the more alcohol consumed, the lower the risk of developing rheumatoid arthritis. Among those who drank regularly, the one-quarter who drank the most were up to 50% less likely to develop the disease compared to the half who drank the least. The researchers noted that moderate alcohol consumption also reduces the risk of other inflammatory processes such as cardiovascualar disease.

Some of the biological mechanisms by which ethanol reduces the risk of destructive arthritis and prevents the loss of bone mineral density (BMD), which is part of the disease process.

A study concluded, "Alcohol either protects from RA or, subjects with RA curtail their drinking after the manifestation of RA". Another study found, "Postmenopausal women who averaged more than 14 alcoholic drinks per week had a reduced risk of rheumatoid arthritis..."

Osteoporosis

Moderate alcohol consumption is associated with higher bone mineral density in postmenopausal women. "...Alcohol consumption significantly decreased the likelihood [of osteoporosis]." "Moderate alcohol intake was associated with higher BMD in postmenopausal elderly women." "Social drinking is associated with higher bone mineral density in men and women [over 45]." However, alcohol abuse is associated with bone loss.

Chronic excessive alcohol abuse is associated with a wide range of skin disorders including urticaria, porphyria cutanea tarda, flushing, cutaneous stigmata of cirrhosis, psoriasis, pruritus, seborrheic dermatitis and rosacea.

A 2010 study concluded, "Nonlight beer intake is associated with an increased risk of developing psoriasis among women. Other alcoholic beverages did not increase the risk of psoriasis in this study."

Immune System, Bacterial Contamination, Viral Infections, and Cancer

There is a protective effect of alcohol consumption against active infection with *H. pylori* In contrast, alcohol intake (comparing those who drink > 30g of alcohol per day to non-drinkers) is not associated with higher risk of duodenal ulcer. Excessive alcohol consumption seen in alcoholics is a known risk factor for pneumonia.

A study on the common cold found that "Greater numbers of alcoholic drinks (up to three or four per day) were associated with decreased risk for developing colds because drinking was associated with decreased illness following infection. However, the benefits of drinking occurred only among nonsmokers. [...] Although alcohol consumption did not influence risk of clinical illness for smokers, moderate alcohol consumption was associated with decreased risk for nonsmokers." Another study concluded, "Findings suggest that wine intake, especially red wine, may have a protective effect against common cold. Beer, spirits, and total alcohol intakes do not seem to affect the incidence of common cold." The International Agency for Research on Cancer (Centre International de Recherche sur le Cancer) of the World Health Organisation has classified alcohol as a Group 1 carcinogen. Its evaluation states, "There is sufficient evidence for the carcinogenicity of alcoholic beverages in humans.... Alcoholic beverages are carcinogenic to humans (Group 1)."

The U.S. Department of Health & Human Services' National Toxicology Program listed alcohol as a *known carcinogen* in 2000.

One study determined that "3.6% of all cancer cases worldwide are related to alcohol drinking, resulting in 3.5% of all cancer deaths." A 2011 study found that one in 10 of all cancers in men and one in 33 in women were caused by past or current alcohol intake. The World Cancer Research Fund panel report *Food, Nutrition, Physical Activity and the*

Prevention of Cancer: a Global Perspective finds the evidence "convincing" that alcoholic drinks increase the risk of the following cancers: mouth, pharynx and larynx, oesophagus, colorectum (men), breast (pre- and postmenopause).

High concentrations of acetaldehyde, which is produced as the body breaks down ethanol, may damage DNA in healthy cells. The National Institute on Alcohol Abuse and Alcoholism have shown that acetaldehyde reacts with polyamines which are naturally occurring compounds essential for cell growth—to create a particularly dangerous type of mutagenic DNA base called a Cr-Pdg adduct. Even moderate levels of alcohol consumption are associated with an increased risk of certain forms of cancer.

Fetal alcohol syndrome or FAS is a disorder of permanent birth defects that occurs in the offspring of women who drink alcohol during pregnancy.

Drinking heavily or during the early stages of prenatal development has been conclusively linked to FAS; moderate consumption is associated with fetal damage. Alcohol crosses the placental barrier and can stunt fetal growth or weight, create distinctive facial stigmata, damaged neurons and brain structures, and cause other physical, mental, or behavioural problems. Fetal alcohol exposure is the leading known cause of mental retardation in the Western world. Alcohol consumption during pregnancy is associated with brain insulin and insulin-like growth factor resistance. Alcohol has been found outside the Solar System. It can be found in low densities in star and planetary system forming regions of space.

Hydrogen Halides

The halogens all form binary compounds with hydrogen known as the hydrogen halides: hydrogen fluoride (HF), hydrogen chloride (HCl), hydrogen bromide (HBr), hydrogen iodide (HI), and hydrogen astatide (HAt). All of these are strong chemical acids when dissolved in water, with the exception of HF. However, hydrofluoric acid does have quite destructive properties towards animal tissues, including those of human beings. When in aqueous solution, the hydrogen halides are known as hydrohalic acids. The names of these acids are as follows: hydrofluoric acid, hydrochloric acid, hydrobromic acid, and hydroiodic acid. All of these acids must be handled with great care because they are dangerous. Some of these acids are also widely used in chemical manufacturing plants.

Hydrogen astatide should also be a strong acid (hydroastatic acid), but it is seldom included in presentations about hydrohalic acids because of the extreme radioactivity of astatine (via alpha decay) and the fact that it readily decomposes into its constituent elements (hydrogen and astatine).

Precursor to Inorganic Compounds

Metal halides are often readily available precursors for other inorganic compounds. Mentioned above, the halide compounds can be made anhydrous by heat, vacuum, or treatment with thionyl chloride.

Halide ligands may be abstracted by silver(I), often as the tetrafluoroborate or the hexafluorophosphate.

In many transition metal compounds, the empty coordination site is stabilized by a coordinating solvent like tetrahydrofuran. Halide ligands may also be displaced by the alkali salt of an X-type ligand, such as a salen-type ligand.

This reaction is formally a transmetallation, and the abstraction of the halide is driven by the precipitation of the resultant alkali halide in an organic solvent. The alkali halides generally have very high lattice energies.

For example, sodium cyclopentadienide reacts with ferrous chloride to yield ferrocene:

$$2\ NaC_5H_5 + FeCl_2 \rightarrow Fe(C_5H_5)_2 + 2\ NaCl$$

While inorganic compounds used for catalysis may be prepared and isolated, they may at times be generated *in situ* by addition of the metal halide and the desired ligand. For example, palladium chloride and triphenylphosphine may be often be used in lieu of bis(triphenylphosphine)palladium(II) chloride for palladium-catalysed coupling reactions.

Interhalogen Compounds

The halogens react with each other to form interhalogen compounds. Diatomic interhalogen compounds such as BrF, ICl, and ClF bear resemblance to the pure halogens in some respects.

The properties and behaviour of a diatomic interhalogen compound tend to be intermediate between those of its parent halogens. Some properties, however, are found in neither parent halogen. For example, Cl_2 and I_2 are soluble in CCl_4, but ICl is not since it is a polar molecule due to the relatively large electronegativity difference between I and Cl.

The halogens react with each other to form interhalogen compounds.

The general formula of most interhalogen compounds is XY_n, where $n = 1, 3, 5$ or 7, and X is the less electronegative of the two halogens. They are all prone to hydrolysis, and ionise to give rise to polyatomic ions. No ternary (containing three different halogens) interhalogen compounds are known.

Organohalogen Compounds

Many synthetic organic compounds such as plastic polymers, and a few natural ones, contain halogen atoms; these are known as *halogenated* compounds or organic halides.

Chlorine is by far the most abundant of the halogens, and the only one needed in relatively large amounts (as chloride ions) by humans. For example, chloride ions play a key role in brain function by mediating the action of the inhibitory transmitter GABA and are also used by the body to produce stomach acid. Iodine is needed in trace amounts for the production of thyroid hormones such as thyroxine.

On the other hand, neither fluorine nor bromine are believed to be essential for humans. Organohalogens are also synthesized through the nucleophilic abstraction reaction.

Polyhalogenated Compounds

Polyhalogenated compounds are industrially created compounds substituted with multiple halogens. Many of them are very toxic and bioaccumulate in humans, and have a very wide application range. They include the much maligned PCBs, PBDEs, and PFCs as well as numerous other compounds.

Production

The element astatine is produced artificially nearly all the time, and only minute amounts are found in nature. Ununseptium is always produced artificially. Note that it is possible that ununseptium may not be a halogen. Nearly all the facts about ununseptium are predicted.

Drug Discovery

In drug discovery, the incorporation of halogen atoms into a lead drug candidate results in analogues that are usually more lipophilic and less water soluble. Consequently, halogen atoms are used to improve penetration through lipid membranes and tissues. It follows that there is a tendency for some halogenated drugs to accumulate in adipose tissue. The chemical reactivity of halogen atoms depends on both their point of attachment to the lead and the nature of the halogen. Aromatic halogen groups are far less reactive than aliphatic halogen groups, which can exhibit considerable chemical reactivity. For aliphatic carbon-halogen bonds the C-F bond is the strongest and usually less chemically reactive than aliphatic C-H bonds. The other aliphatic-halogen bonds are weaker, their reactivity increasing down the periodic table. They are usually more chemically reactive than aliphatic C-H bonds. Consequently, the most common halogen substitutions are the less reactive aromatic fluorine and chlorine groups. The halogens chlorine and iodine are needed by our bodies in trace amounts. We eat chlorine in salts, and we eat iodine in vegetables. The other three halogens have no known biological role.

Toxicity

Halogens are very reactive elements and their toxicity derives from their strong oxidant properties. Fluorine is the most toxic and dangerous of the halogens, capable of killing a person within seconds. Chlorine is also toxic, and can kill a person if inhaled in an enclosed area. Bromine is less reactive and toxic than Chlorine however, being a liquid at standard temperature and pressure, it is much more dense yet still behaves as a fuming liquid. Contact with molecular Bromine can be fatal. Bromine's increased reactivity with water can also lead to the formation of a higher quantity of acids in the lungs than those that might be formed with Chlorine. Iodine is the least dangerous of the halogens and will rarely kill a person if its vapor is inhaled however ingestion can be fatal even at a low dose (28 mg/kg).

Chapter 3

Carbonyl Complexes

Dicobalt octacarbonyl reacts with hydrogen and alkenes to give aldehydes. This reaction is the basis of hydroformylation, the formation of aldehydes from an alkene, CO and hydrogen. A key intermediate is cobalt tetracarbonyl hydride ($HCo(CO)_4$). The original Ruhrchemie process produced propanal from ethene and syngas using cobalt carbonyl has been displaced by rhodium-based catalysts. Processes involving cobalt are practiced by BASF, EXXON, and Shell mainly for the production of C7-C14 alcohols used for the production of surfactants.

In hydrocarboxylations hydrogen is replaced by water or an alcohol and the reaction product is a carboxylic acid or an ester. An example of this reaction type is the conversion of butadiene to adipic acid.

Alkyne Derivatives of $Co_2(CO)_8$

Dicobalt octacarbonyl also reacts with alkynes to give "tetrahedranes" of the formula $Co_2(CO)_6(C_2R_2)$.

Because the cobalt carbonyl centres can be removed later, it functions as a protective group for the alkyne. In the Nicholas reaction an alkyne group is also protected and at the same time the alpha-carbon position is activated for nucleophilic substitution.

Cyclization Reactions

Cobalt compounds react with dialkynes and dienes to cyclic intermediates in cyclometalation. Other alkynes, alkenes, nitriles or carbon monoxide can then insert themselves into the Co-C bond. Reaction types based on this concept are the Pauson–

Khand reaction (CO insertion) and alkyne trimerization (notably with cyclopentadienylcobalt dicarbonyl).

Cp, Allyl, and Alkene Compounds

In 1866, Berthelot reported the first example of cyclotrimerization leading to aromatic products, the cyclization of acetylene to benzene. The reaction required very high temperatures to proceed (400 °C) and produced a complex mixture of products, and research in the field remained dormant until the late 1940s. In 1948, Reppe discovered that nickel could catalyse the formation of substituted benzenes from acetylenic compounds; since his initial discovery, cyclotrimerizations to produce substituted benzenes have been catalysed by no less than seventeen transition metals, including:

- Group 4 (Ti, Zr)
- Group 5 (V, Nb)
- Group 6 (Mo, W)
- Group 8 (Fe, Ru, Os)
- Group 9 (Co, Rh, Ir)
- Group 10 (Pd)
- Group 11 (Cu)

Advantages: Control of the substitution pattern of the aromatic product is good in many cases, and cyclotrimerization can be used in cases when functionalization of pre-formed aromatic materials (through electrophilic aromatic substitution, for instance) is not selective. The reaction is highly chemoselective for triple bonds and can tolerate a wide variety of functional groups on the starting materials. Disadvantages: For cotrimerizations involving two or three different acetylenes, a variety of regioisomers may form. Operationally, these reactions usually require elevated temperatures (>60 °C) and sometimes require irradiation to facilitate the dissociation of strongly binding carbon monoxide ligands.

Mechanism and Stereochemistry

Prevailing Mechanism

The most common mechanism for the cyclotrimerization of acetylenes begins with the formation of a metallo cyclo pentadiene complex. Oxidative cyclization of two coordinated alkyne units produces either metallocycle 3 or 4. After dissociation of a third ligand and coordination of a third alkyne, two pathways are possible. Either alkyne insertion generates metallocyclohep-tatriene 5, or [4+2] cycloaddition generates bridged bicycle 6.

Enantioselective Variants

Chiral catalysts have been employed in combination with arynes to produce non-racemic amounts of atropisomeric products. Selective cyclotrimerization of one of a pair of enantiotopic alkynes has also been facilitated by a chiral catalyst.

Scope and Limitations

Catalysts for cyclotrimerization are highly selective for triple bonds, which gives the reaction a fairly wide substrate scope. Alcohols, ethers, amines, and carbonyl compounds (ketones, esters, amides, and carboxylic acids) are well tolerated.

Other than carbonyl compounds, the reaction is not generally compatible with unsaturated functionality. Nitriles, for instance, can react to form pyridines.

The reaction is limited in some cases by catalyst deactivation due to the formation of stable, 18-electron η^4-complexes. Cyclobutadiene, cyclohexadiene, and arene complexes have all been observed as off-cycle, inactivate catalyst forms.

Control via Degenerative Transfer

One of the disadvantages of controlled radical polymerization reactions is that they tend to become rather slow. Controlled polymerization conditions are usually achieved by extending the life-time of the growing polymer chain radical, by keeping it in a dormant state for most of the time (known as the Persistent Radical Effect).

Thereby the control agent substantially slows-down the over-all radical polymerisation reaction. However, some CMRP reactions proceed via a different mechanism, called degenerative transfer (DT), which allows controlled radical polymerization reactions to proceed at roughly the same rate as any uncontrolled free radical polymerisation.

The degenerative transfer mechanism is based on very fast exchange equilibria between small free radicals (being continuously injected into the solution) and dormant polymeryl radicals (protected as closed-shell cobalt species).

Systems based on degenerative transfer do not proceed via the persistent radical effect (PRE). Instead an active propagating radical interchanges its role with a latent radical in a dormant complex. The activation of one polymer chain means the deactivation of another polymer chain. If the exchange process is much faster than the polymerisation rate (k_p), effectively all polymer chains grow at the same rate. Because the large polymer chains diffuse much slower than the small organic radicals, and thereby terminate much slower via 2nd order radical-radical coupling or disproportionation, long chains effectively build-up at cobalt while the small radicals keep terminating.

This leads to a desirable narrow molecular weight distribution of the polymer at high polymerization rates. DT-CMRP is an associative process, which for Co^{III}(por)(alkyl) species implies formation of a 6-coordinate intermediate or transition state. Such Co(por)$(alkyl)_2$ species formally have a Co(+IV) oxidation state, but in reality their (electronic) structure is best described as a weak radical adduct of a Co^{III}(por)(alkyl) species. A striking feature of DT-CMRP is the fact that even upon using a large excess of the radical initiator compared to the transfer agent, the radical polymerization reactions still remains controlled. A satisfactory explanation for this phenomenon seems to be lacking at the moment.

Fischer-Tropsch Catalysis

Cobalt catalysts (together with iron) are relevant in the Fischer-Tropsch process in which synthesis gas is converted to hydrocarbons. In this process, it is assumed that organocobalt intermediates form. An idealized reaction sequence is depicted below :

$$M + CO \rightarrow M\text{-}CO \ (M = Co, Fe)$$

$$M\text{-}CO + H_2 \rightarrow M\text{-}CH_3$$

$$M\text{-}CH_3 + CO \rightarrow OC\text{-}M\text{-}CH_3$$

$$OC\text{-}M\text{-}CH_3 \rightarrow M\text{-}(CO)\text{-}CH_3$$

$$M\text{-}(CO)\text{-}CH_3 + H_2 \rightarrow M\text{-}CH_2CH_3$$

Radical Polymerization

Free radical polymerization is a method of polymerization by which a polymer forms by the successive addition of free radical building blocks. Free radicals can be formed via a number of different mechanisms usually involving separate initiator molecules. Following its generation, the initiating free radical adds (nonradical) monomer units, thereby growing the polymer chain.

Free radical polymerization is a key synthesis route for obtaining a wide variety of different polymers and material composites. The relatively non-specific nature of free radical chemical interactions makes this one of the most versatile forms of polymerization available and allows facile reactions of polymeric free radical chain ends and other chemicals or substrates. In 2001, 40 billion of the 110 billion pounds of polymers produced in the United States were produced by free radical polymerization.

Free radical polymerization is a type of chain growth polymerization, along with anionic, cationic and coordination polymerization. Initiation is the first step of the polymerization process. During initiation, an active centre is created from which a polymer chain is generated. Not all monomers are susceptible to all types of initiators. Radical initiation works best on the carbon-carbon double bond of vinyl monomers and the carbon-oxygen double bond in aldehydes and ketones. Initiation has two steps. In the first step, one or two radicals are created from the initiating molecules. In the second step, radicals are transferred from the initiator molecules to the monomer units present.

Thermodynamics

In chain growth polymerization, the position of the equilibrium between polymer and monomers can be determined by the thermodynamics of the polymerization. The Gibbs free energy (ΔG_p) of the polymerization is commonly used to quantify the tendency of a polymeric reaction. The polymerization will be favoured if $\Delta G_p < 0$; if $\Delta G_p > 0$, the polymer will undergo depolymerization. According to the thermodynamic equation $\Delta G = \Delta H - T\Delta S$, a negative enthalpy and an increasing entropy will shift the equilibrium towards polymerization.

In general, the polymerization is an exothermic process, i.e. negative enthalpy change, since addition of a monomer to the growing polymer chain involves the conversion of π bonds into σ bonds or an ring opening reaction that releases the ring tension in a cyclic monomer. Meanwhile, during polymerization, a large amount of small molecules are associated, losing rotation and translational degrees of freedom. As a result, the entropy decreases in the system, $\Delta S_p < 0$ for nearly all polymerization processes. Since depolymerization is almost always entropically favoured, the ΔH_p must then be sufficiently negative to composite for the unfavourable entropic term. Only then will polymerization be thermodynamically favoured by the resulting negative ΔG_p.

In practice, polymerization is favoured at low temperatures: $T\Delta S_p$ is small. Depolymerization is favoured at high temperatures: $T\Delta S_p$ is large. As the temperature increases, ΔG_p become less negative. At certain temperature, the polymerization reaches equilibrium (rate of polymerization = rate of depolymerization). This temperature is called the ceiling temperature (T_c). $\Delta G_p = 0$

Stereochemistry of Polymerization

The stereochemistry of polymerization is concerned with the difference in atom connectivity and spatial orientation in polymers that has the same chemical composition. Staudinger studied the stereoisomerism in chain polymerization of vinyl monomers in late 1920s, and it took another two decades for people to fully appreciate the idea that each of the propagation steps in the polymer growth could give rise to stereoisomerism.

The major milestone in the stereochemistry was established by Ziegler and Natta and their coworkers in 1950s, as they developed metal based catalyst to synthesize stereoregular polymers.

The reason why the stereochemistry of the polymer is of particular interest is because the physical behaviour of a polymer depends not only on the general chemical composition but also on the more subtle differences in microstructure. Atactic polymers consist of a random arrangement of stereochemistry and are amorphous (noncrystalline), soft materials with lower physical strength.

The corresponding isotactic (like substituents all on the same side) and syndiotactic (like substituents of alternate repeating units on the same side) polymers are usually obtained as highly crystalline materials.

It is easier for the stereoregular polymers to pack into a crystal lattice since they are more ordered and the resulting crystallinity leads to higher physical strength and increased solvent and chemical resistance as well as differences in other properties that depend on crystallinity.

The prime example of the industrial utility of stereoregular polymers is polypropene. Isotactic polypropene is a high-melting (165 °C), strong, crystalline polymer, which is used as both a plastic and fibre. Atactic polypropene is an amorphous material with an oily to waxy soft appearance that finds use in asphalt blends and formulations for lubricants, sealants, and adhesives, but the volumes are minuscule compared to that of isotactic polypropene.

When a monomer adds to a radical chain end, there are two factors to consider regarding its stereochemistry: 1) the interaction between the terminal chain carbon and the approaching monomer molecule and 2) the configuration of the penultimate repeating unit in the polymer chain. The terminal carbon atom has sp^2 hybridization and is planar. Consider the polymerization of the monomer CH_2=CXY. There are two ways that a monomer molecule can approach the terminal carbon: the mirror approach (with like substituents on the same side) or the non-mirror approach (like substituents on opposite sides).

Reactivity

Traditionally, the reactivity of monomers and radicals are assessed by the means of copolymerization data. *Q-e* scheme, the most widely used tools for the semiquantitative prediction of monomer reactivity ratios, was first proposed by Alfrey and Price in 1940s. The scheme takes into account the intrinsic thermodynamic stability and polar effects in the transition state. A given radical and a monomer is considered to have an intrinsic reactivity of Q_1 and Q_2, respectively. The polar effects in the transition state, the supposed permanent electric charge carried by that entity (radical or molecule), is quantified by the factor *e*, which is a constant for a given monomer, and has the same value for the radical derived from that specific monomer. For reaction between a radical (species 1) and a monomer (species 2), the rate constant, k_{12}, was postulated to be related to the four relevant reactivity parameters by

$$k_{12} = Q_1 Q_2 exp(-e_1 e_2)$$

The monomer reactivity ratios for the copolymerization of monomers 1 and 2 can be given by

Free radical polymerization has found myriad applications including, but not limited to, the manufacture of polystyrene, thermoplastic block copolymer elastomers (which may be used for a wide variety of applications including adhesives, footwear, and toys), cardiovascular stents, chemical surfactants, and lubricants. Free radical polymerization has many uses in research as well. One novel and particularly interesting application, one that exemplifies the power of the technique, is its use in the functionalization of carbon nanotubes. Carbon nanotubes due to their intrinsic electronic properties, tend to form large aggregates in solution, precluding their use for useful applications. Adding small chemical groups to the walls of nanotubes can eliminate this propensity toward aggregation and can be used to tune the response of a nanotube to its surrounding environment; the use of polymers instead of smaller molecules can be used to drastically modify nanotube properties (and conversely, nanotubes can be used to modify polymer mechanical and electronic properties). For example, Lou et al. were able to demonstrate the coating of carbon nanotubes by polystyrene by first polymerizing polystyrene via chain radical polymerization and subsequently mixing it at 130 °C with carbon nanotubes to generate polystyrene radicals and graft them onto the walls of carbon nanotubes. The advantage of this approach lies in the order of chemical reaction – rather than growing a polymer off of a carbon nanotube (the "grafting from" approach), chain growth polymerization is used to first synthesize a polymer with predetermined properties. Synthesis of these materials is currently possible only by the use of free radical polymerization. The clay is first dispersed in water where it forms very small, porous plates. Subsequent addition of the organic monomer, generally an acrylamide or acrylamide derivative, is immediately proceeded by addition of the initiator and a catalyst. The initiator is chosen to have stronger interaction with the clay than the organic monomer, so it preferentially adsorbs to the surface of the clay.

The entire mixture of clay, organic monomer, initiator, catalyst, and water solvent is heated to initiate polymerization. Polymers grow off of the initiators which are in turn bound to the clay.

Due to recombination and disproportionation reactions, growing polymer chains bind to one another, forming a strong, cross-linked network polymer, with clay particles acting as branching points for multiple polymer chain segments.

Living Polymerization

In polymer chemistry, living polymerization is a form of addition polymerization where the ability of a growing polymer chain to terminate has been removed. This can be accomplished in a variety of ways.

Chain termination and chain transfer reactions are absent and the rate of chain initiation is also much larger than the rate of chain propagation. The result is that the polymer chains grow at a more constant rate than seen in traditional chain polymerization and their lengths remain very similar (i.e. they have a very low polydispersity index).

Living polymerization is a popular method for synthesizing block copolymers since the polymer can be synthesized in stages, each stage containing a different monomer. Additional advantages are predetermined molar mass and control over end-groups.

Living polymerization in the literature is often called "living" polymerization or controlled polymerization. Living polymerization was demonstrated by Michael Szwarc in 1956 in the anionic polymerization of styrene with an alkali metal / naphthalene system in tetrahydrofuran (THF).

He found that after addition of monomer to the initiator system that the increase in viscosity would eventually cease but that after addition of a new amount of monomer after some time the viscosity would start to increase again.

The main living polymerization techniques are:

- Living anionic polymerization
- Living cationic polymerization
- Ring opening metathesis polymerization
- Living free radical polymerization
- Group transfer polymerization
- living Ziegler-Natta polymerization

Living Anionic Polymerization

As carly as 1936, Karl Ziegler proposed that anionic polymerization of styrene and butadiene by consecutive addition of monomer to an alkyl lithium initiator occurred without chain transfer or termination. Twenty years later, living polymerization was demonstrated by Szwarc through the anionic polymerization of styrene in THF using sodium naphthalenide as celerator. Living anionic polymerization is living polymerization technique involving an anionic propagating species.

Living anionic polymerization was demonstrated by Szwarc and co workers in 1956. Their initial work was based on the polymerization of styrene and dienes. One of the remarkable features of living anionic polymerization is that the mechanism involves no formal termination step. In the absence of impurities, the carbanion would still be active and capable of adding another monomer. The chains will remain active indefinitely unless there is inadvertent or deliberate termination or chain transfer. This gave rise to two important consequences:

1. The number average molecular weight, Mn, of the polymer resulting from such a system could be calculated by the amount of consumed monomer and the initiator used for the polymerization, as the degree of polymerization would be the ratio of the moles of the monomer consumed to the moles of the initiator added.

   $M_n = M_o \frac{[M]_o}{[I]}$, where M_o = formula weight of the repeating unit, $[M]_0$ = initial concentration of the monomer, and [I] = concentration of the initiator.

2. All the chains are initiated at roughly the same time. The final result is that the polymer synthesis can be done in a much more controlled manner in terms of the molecular weight and molecular weight distribution (Poisson distribution).

The following experimental criteria have been proposed as a tool for identifying a system as living polymerization system.

- Polymerization until the monomer is completely consumed and until further monomer is added.
- Constant number of active centres or propagating species.
- Poisson distribution of molecular weight
- Chain end functionalization can be carried out quantitatively.

However, in practice, even in the absence of terminating agents, the concentration of the living anions will reduce with time due to a decay mechanism termed as spontaneous termination.

Synthesis of Complex Architectures

Polymerization reactions excluding a termination and transfer step are particularly useful for the synthesis of functionalized polymers with well defined architectures.. The consumption of the monomer results in stable, anionic polymer chain ends, allowing reactions with a variety of electrophilic functional groups post polymerization.

$$PLi + X\text{-}Y \rightarrow P\text{-}X + LiY$$

However the efficacy of these reactions depends on a number of variables such as chain-end structure, solvent, temperature, and concentration. Alternatively, by controlling the functionality of the initiator, one can prepare polymers having different geometries such as symmetric and asymmetric stars, comb shaped, etc.

Living Cationic Polymerization

Monomers for living cationic polymerization are electron-rich alkenes such as vinyl ethers, isobutylene, styrene, and N-vinylcarbazole. The initiators are binary systems consisting of a electrophile and a Lewis acid. The method was developed around 1980 with contributions from Higashimura, Sawamoto and Kennedy.

Living cationic polymerization is a living polymerization technique involving cationic propagating species.

It enables the synthesis of very well defined polymers (low molar mass distribution) and of polymers with unusual architecture such as star polymers and block copolymers and living cationic polymerization is therefore as such of commercial and academic interest.

The basic reaction steps are: iodine. In carbocationic polymerization the active site is a carbocation with a counterion in close proximity.

$A^+B^- + H_2C{=}CHR \rightarrow A\text{-}CH_2\text{-}RHC^+\text{——}B^-$

- Chain propagation:

$A\text{-}CH_2\text{-}RHC^+\text{——}B^- + H_2C{=}CHR \rightarrow A\text{-}(CH_2\text{-}RHC)_n\text{-}CH_2\text{-}RHC^+\text{——}B^-$

- Chain termination:

$A\text{-}(CH_2\text{-}RHC)_n\text{-}CH_2\text{-}RHC^+\text{——}B^- \rightarrow A\text{-}(CH_2\text{-}RHC)_n\text{-}CH_2\text{-}RHC\text{-}B$

- chain transfer:

$A\text{-}(CH_2\text{-}RHC)_n\text{-}CH_2\text{-}RHC^+\text{——}B^- \rightarrow A\text{-}(CH_2\text{-}RHC)_n\text{-}CH_2{=}CR\ H^+B^-$

Living cationic polymerization is characterised by defined and controlled initiation and propagation while minimising side-reactions termination and chain transfer.

Transfer and termination do occur but in ideal living systems the active ionic propagating species are in chemical equilibrium with the dormant covalent species with an exchange rate much faster than the propagation rate.

Solution methods require rigorous purification of monomer and solvent although conditions are not as strict as in anionic polymerization.

Common monomers are vinyl ethers, alpha-methyl vinyl ethers, isobutene, styrene, methylstyrene and N-vinylcarbazole. The monomer is nucleophilic and substituents should be able to stabilize a positive carbocationic charge. For example para-methoxystyrene is more reactive than styrene itself.

Initiation takes place by a initiation/coinitiation binary system, for example an alcohol and a Lewis acid. The active electrophile is then a proton and the counter ion the remaining alkoxide which is stabilized by the Lewis acid. With organic acetates such as cumyl acetate the initiating species is the carbocation R^+ and the counterion is the acetate anion.

In the iodine/HI system the electrophile is again a proton and the carbocation is stabilized by the triiodide ion. Polymerizations with Diethylaluminium chloride rely on trace amounts of water. A proton is then accompanied by the counterion $Et_2AlClOH^-$. With tert-butyl chloride Et_2AlCl abstracts a chlorine atom to form the tert-butyl carbocation as the electrophile. Efficient initiators that resemble the monomer are called cationogens. Termination and chain transfer are minimised when the initiator counterion is both non-nucleophilic and non-basic. More polar solvents promote ion dissociation and hence increase molar mass.

Common additives are electron donors, salts and proton traps . Electron donors (e.g. nucleophiles, Lewis bases) for example dimethylsulfide and dimethylsulfoxide are believed to stabilize the carbocation.

The addition of salt for example a tetraalkylammonium salt, prevents dissociation of the ion pair that is the propagating reactive site. Ion dissociation into free ions lead to non-living polymerization. Proton traps scavenge protons originating from protic impurities. The method was developed starting in the 1970s and 1980s with contributions from Higashimura on the polymerization of p-methoxystyrene using iodine or acetyl perchlorate, on the polymerization of isobutyl vinyl ether by iodine and with Mitsuo Sawamoto by iodine/HI and on the formation of p-methoxystyrene - isobutyl vinyl ether block copolymers. Kennedy and Faust studied methylstyrene / boron trichloride polymerization (then called quasi-living) in 1982 and that of isobutylene (system with cumyl acetate, 2,4,4-trimethylpentane-2-acetate and BCl_3) in 1984.

Isobutylene Polymerization

Living isobutylene polymerization typically takes place in chloroform or dichloromethane as a solvent at temperatures below 0°C. With more polar solvents polyisobutylene solubility becomes a problem. Initiators can be esters, alcohols, halides and ethers. Co-initiators are boron trichloride, tin tetrachloride and organoaluminium halides. With ethers and alcohols the true initiator is the chlorinated product. Polymer with molar mass of 160,000 g/mole and polydispersity index 1.02 can be obtained.

Vinyl Ether Polymerization

Vinyl ethers (CH_2=CHOR, R = methyl, ethyl, isobutyl, benzyl) are very reactive vinyl monomers. Studied systems are based on I_2/HI and on zinc halides zinc chloride, zinc bromide and zinc iodide.

Living Cationic Ring-Opening Polymerization

In Living cationic ring-opening polymerization the monomer is a heterocycle such as an epoxide, THF, an oxazoline or an aziridine such as t-butylaziridine. The propagating species is not a carbocation but an oxonium ion.

Living polymerization is more difficult to achieve because of the ease of termination by nucleophilic attack of a heteroatom in the growing polymer chain. Intramolecular termination is called backbiting and results in the formation of cyclic oligomers. Initiators are strong electrophiles such as triflic acid. Triflic anhydride is an initiator for bifunctional polymer.

Living Ring-Opening Metathesis Polymerization

Given the right reaction conditions ring-opening metathesis polymerization (ROMP) can be rendered living. The first such systems were described by Robert H. Grubbs in 1986 based on norbornene and Tebbe's reagent and in 1978 Grubbs together with Richard R. Schrock describing living polymerization with a tungsten carbene complex.

Living Free Radical Polymerization

Starting in the 1970s several new methods were discovered which allowed the development of living polymerization using free radical chemistry.

These techniques involved catalytic chain transfer polymerization, iniferter mediated polymerization, stable free radical mediated polymerization (SFRP), atom transfer radical polymerization (ATRP), reversible addition-fragmentation chain transfer (RAFT) polymerization, and iodine-transfer polymerization.

Living free radical polymerization is a type of living polymerization where the active polymer chain end is a free radical. Several methods exist. IUPAC recommends to use the term "reversible-deactivation radical polymerization" instead of "living free radical polymerization".

Catalytic Chain Transfer and Cobalt Mediated Radical Polymerization

Although not a strictly living form of polymerization catalytic chain transfer polymerization must be mentioned as it figures significantly in the development of later forms of living free radical polymerization.

Discovered in the late 1970s in the USSR it was found that cobalt porphyrins where able to reduce the molecular weight during polymerization of methacrylates. Later investigations showed that the cobalt glyoxime complexes were as effective as the porphyrin catalysts and also less oxygen sensitive. Due to their lower oxygen sensitivity these catalysts have been investigated much more thoroughly than the porphyrin catalysts. The major products of catalytic chain transfer polymerization are vinyl terminated polymer chains. One of the major drawbacks of the process is that catalytic chain transfer polymerization does not produce macromonomers but instead produces addition fragmentation agents. When a growing polymer chain reacts with the addition fragmentation agent the radical end-group attacks the vinyl bond and forms a bond. However, the resulting product is so hindered that the species undergoes fragmentation, leading eventually to telechelic species. These addition fragmentation chain transfer agents do form graft copolymers with styrenic and acrylate species however they do so by first forming block copolymers and then incorporating these block copolymers into the main polymer backbone. While high yields of macromonomers are possible with methacrylate monomers, low yields are obtained when using catalytic chain transfer agents during the polymerization of acrylate and stryenic monomers. This has been seen to be due to the interaction of the radical centre with the catalyst during these polymerization reactions.

The reversible reaction of the cobalt macrocycle with the growing radical is known as cobalt carbon bonding and in some cases leads to living polymerization reactions.

Catalytic chain transfer (CCT) is a process that can be incorporated into radical polymerization to obtain greater control over the resulting products.

Radical polymerization of vinyl monomers, like methyl (metha) acrylate of vinyl acetate is a common (industrial) method to prepare polymeric materials. One of the problems associated with this method is, however, that the radical polymerisation reaction rate is so high that even at short reaction times the polymeric chains are exceedingly long. This has several practical disadvantages, especially for polymer processing (e.g. melt-processing). A solution to this problem is catalytic chain transfer, which is a way to make shorter polymer chains in radical polymerisation processes. The method involves adding a catalytic chain transfer agent to the reaction mixture of the monomer and the radical initiator.

Boris Smirnov and Alexander Marchenko (USSR) discovered in 1975 that cobalt porphyrins are able to reduce the molecular weight of PMMA formed during radical polymerization of methacrylates. Later investigations showed that the cobalt dimethylglyoxime complexes were as effective as the porphyrin catalysts and also less oxygen sensitive. Due to their lower oxygen sensitivity these catalysts have been investigated much more thoroughly than the porphyrin catalysts and are the catalysts actually used commercially.

Utility

The catalytic chain transfer process was commercialised very soon after its discovery. The initial commercial outlet was the production of chemically-reactive macromonomers to be incorporated into paints for the automotive industry. Federally-mandated VOC restrictions are leading to the elimination of solvents from the automotive finishes and the lower molecular weight chain transfer products are often fluids. Incorporation of monomers such as glycidyl methacrylate or hydroxyethylmethacrylate (HEMA) into the macromonomers aid curing processes. Macromonomers incorporating HEMA can be effective in the dispersion of pigments in the paints.

Components of ATRP

There are five important variable components of Atom Transfer Radical Polymerizations. They are the monomer, initiator, catalyst, solvent and temperature. The following section breaks down the contributions of each component to the overall polymerization.

Monomer

Monomers that are typically used in ATRP are molecules with substituents that can stabilize the propagating radicals; for example, styrenes, (meth)acrylates, (meth)acrylamides, and acrylonitrile.

ATRP are successful at leading to polymers of high number average molecular weight and a narrow polydispersity index when the concentration of the propagating radical balances the rate of radical termination.

Yet, the propagating rate is unique to each individual monomer. Therefore, it is important that the other components of the polymerization (initiator, catalysts, ligands and solvents) are optimized in order for the concentration of the dormant species to be greater than the concentration of the propagating radical and yet not too great to slow down or halt the reaction.

Catalyst

The catalyst is the most important component of ATRP because it determines the equilibrium constant between the active and dormant species. This equilibrium determines the polymerization rate and an equilibrium constant too small may inhibit or slow the polymerization while an equilibrium constant too large leads to a high distribution of chain lengths.

There are several requirements for the metal catalyst:

1. there needs to be two accessible oxidation states that are separated by one electron
2. the metal centre needs to have a reasonable affinity for halogens
3. the coordination sphere of the metal needs to be expandable when its oxidized so to be able to accommodate the halogen
4. a strong ligand complexation.

The most studied catalysts are those that polymerizations involving copper, which has shown the most versatility, showing successful polymerizations regardless of the monomer.

Solvent

Toluene,1,4-dioxane, xylene, anisole, DMF, DMSO, water, methanol, acetonitrile, chloroform, bulk monomer

Temperature

Reverse ATRP

In reverse ATRP, the catalyst is added in its higher oxidation state. Chains are activated by conventional radical initiators (e.g. AIBN) and deactivated by the transition metal. The source of transferrable halogen is the copper salt, so this must be present in concentrations comparable to the transition metal. A mixture of radical initiator and active (lower oxidation state) catalyst allows for the creation of block copolymers (contaminated with homopolymer) which is impossible using standard reverse ATRP. This is called SR&NI (simultaneous reverse and normal initiation ATRP).

AGET ATRP

Activators generated by electron transfer uses a reducing agent unable to initiate new chains (instead of organic radicals) as regenerator for the low-valent metal. Examples are metallic Cu, tin(II), ascorbic acid, or triethylamine. It allows for lower concentrations of transition metals, and may also be possible in aqueous or dispersed media.

Hybrid and Bimetallic Systems

This technique uses a variety of different metals/oxidation states, possibly on solid supports, to act as activators/ deactivators, possibly with reduced toxicity or sensitivity. Iron salts can, for example, efficiently activate alkyl halides but requires an efficient Cu(II) deactivator which can be present in much lower concentrations (3–5 mol%)

ICAR ATRP

Initiators for continuous activator regeneration is a technique that uses large excesses of initiator to continuously regenerate the activator, lowering its required concentration from thousands of ppm to around 1 ppm; making it an industrially relevant technique. Styrene is especially interesting because it generates radicals when sufficiently heated.

ARGET ATRP

Activators regenerated by electron transfer can be used to make block copolymers using a method similar to AGET but requiring strongly reduced amounts of metal, since the activator is regenerated from the deactivator by a large excess of reducing agent (e.g. hydrazine, phenoles, sugars, ascorbic acid, etc...) It differs from AGET ATRP in that AGET uses reducing agents to generate the active catalyst (in quasi stoichiometric amounts) while in ARGET a large excess is used to continuously regenerate the activator allowing transition metal concentrations to drop to ~1 ppm without loss of control.

The calculation of molecular weight for a synthesized polymer is relatively easy, in spite of the complex mechanism for RAFT polymerization. As stated before, during the equilibration step, all chains are growing at equal rates, or in other words, the molecular weight of the polymer increases linearly with conversion. Multiplying the ratio of monomer consumed to the concentration of the CTA used by the molecular weight of the monomer (mM) a reliable estimate of the number average molecular weight can be determined.

Reversible Addition Fragmentation chain Transfer polymerization or RAFT is a degenerative chain transfer process and is free radical in nature. RAFT agents contain di- or tri-thiocarbonyl groups, and it is the reaction with an initiator, usually AIBN, that creates a propagating chain or polymer radical.

This polymer chain then adds to the C=S and leads to the formation of a stabilized radical intermediate. In an ideal system, these stabilized radical intermediates do not undergo termination reactions, but instead reintroduce a radical capable of reinitiation or propagation with monomer, while they themselves reform their C=S bond. The cycle of addition to the C=S bond, followed by fragmentation of a radical, continues until all monomer or intitiator is consumed. Termination is limited in this system by the low concentration of active radicals and any termination that does occur is negligible.

RAFT, invented by Rizzardo *et al.* at CSIRO and a mechanistically identical process termed Macromolecular Design via Interchange of Xanthates (MADIX), invented by Zard *et al.* at Rhodia were both first reported in 1998/early 1999.

Reversible Addition-Fragmentation chain Transfer or RAFT polymerization is one of several kinds of controlled radical polymerization. It makes use of a chain transfer agent in the form of a thiocarbonylthio compound (or similar, from here on referred to as a RAFT agent) to afford control over the generated molecular weight and polydispersity during a free-radical polymerization.

Discovered at the Commonwealth Scientific and Industrial Research Organisation (CSIRO) in 1998, RAFT polymerization is one of several living or Controlled radical polymerization techniques, others being atom transfer radical polymerization (ATRP) and nitroxide-mediated polymerization (NMP).

RAFT polymerization uses thiocarbonylthio compounds, such as dithioesters, thiocarbamates, and xanthates, to mediate the polymerization via a reversible chain-transfer process.

As with other controlled radical polymerization techniques, RAFT polymerizations can be performed with conditions to favour low polydispersity indices and a pre-chosen molecular weight. RAFT polymerization can be used to design polymers of complex architectures, such as linear block copolymers, comb-like, star, brush polymers and dendrimers.

The addition-fragmentation chain transfer process was first reported in the early 1970s. However, the technique was irreversible, so the transfer reagents could not be used to control radical polymerization at this time. For the first few years RAFT was used to help synthesize end-functionalized polymers. Scientists began to realise the potential of RAFT in controlled radical polymerization in the 1980s. Macromonomers were known as reversible chain transfer agents during this time, but had limited applications on controlled radical polymerization. In 1995, a key step in the "degenerate" reversible chain transfer step for chain equilibration was brought to attention. The essential feature is that the product of chain transfer is also a chain transfer agent with similar activity to the precursor transfer agent.

RAFT polymerization today is mainly carried out by thiocarbonylthio chain transfer agents. It was first reported by Thang *et al.* in 1998.

RAFT is one of the most versatile methods of controlled radical polymerization because it is tolerant of a very wide range of functionality in the monomer and solvent, including aqueous solutions. RAFT polymerization has also been effectively carried out over a wide temperature range.

Important Components of RAFT:

Typically, a RAFT polymerization system consists of:

- A radical source (e.g. thermochemical initiator or the interation of gamma radiation with some reagent)
- Monomer
- RAFT agent
- Solvent (not strictly required if the monomer is a liquid)

RAFT Mechanism

RAFT is a type of living polymerization involving a conventional radical polymerization which is mediated by a RAFT agent. Monomers must be capable of radical polymerization. There are a number of steps in a RAFT polymerization: initiation, pre-equilibrium, re-initiation, main equilibrium, propagation and termination.

Initiation: The reaction is started by a free-radical source which may be a decomposing radical initiator such as AIBN. In the example, the initiator decomposes to form two fragments (I•) which react with a single monomer molecule to yield a propagating (i.e. growing) polymeric radical of length 1, denoted P_1•. Propagation: Propagating radical chains of length *n* in their active (radical) form, P_n•, add to monomer, M, to form longer propagating radicals, P_{n+1}•.

RAFT pre-equilibrium: A polymeric radical with n monomer units (P_n) reacts with the RAFT agent to form a RAFT adduct radical. This may undergo a fragmentation reaction in either direction to yield either the starting species or a radical (R•) and a polymeric RAFT agent (S=C(Z)S-P_n). This is a reversible step in which the intermediate RAFT adduct radical is capable of losing either the R group (R•) or the polymeric species (P_n•).

Re-initiation: The leaving group radical (R•) then reacts with another monomer species, starting another active polymer chain.

Main RAFT equilibrium: This is the most important part in the RAFT process, in which, by a process of rapid interchange, the present radicals (and hence opportunities for polymer chain growth) are "shared" among all species that have not yet undergone termination (P_n• and S=C(Z)S-P_n). Ideally the radicals are shared equally, causing chains to have equal opportunities for growth and a narrow PDI.

Termination: Chains in their active form react via a process known as bi-radical termination to form chains that cannot react further, know as dead polymer. Ideally, the RAFT adduct radical is sufficiently hindered such that it does not undergo termination reactions.

Thermodynamics of the Main RAFT Equilibrium

The position of the Main RAFT equilibrium is affected by the relative stabilities of the RAFT adduct radical (P_n-S-C•(Z)-S-P_m) and its fragmentation products, namely S=C(Z)-P_n and polymeric radical (P_m•). If formation of the RAFT adduct radical is sufficiently thermodynamically favourable, the concentration of active species, P_m•, will be reduced to the extent that a reduction in the rate of conversion of monomer into polymer is also observed, as compared to an equivalent polymerization without RAFT agent. Such a polymerization, is referred to as a rate-retarded RAFT polymerization. The rate of a RAFT polymerisation, that is, the rate of conversion of monomer into polymer, depends on the rate of the Propagation reaction, which is proportional to the concentration, [P•], of the active species P•, whereas the rate of the termination reaction, being second order, is proportional to the square $[P\bullet]^2$. This means that during rate-retarded RAFT polymerizations, the rate of formation of termination products is suppressed to a greater extent than the rate of chain growth.

In RAFT polymerisations without rate-retardation, the concentration of the active species P• is close to that in an equivalent conventional polymerisation in the absence of RAFT agent. The main RAFT equilibrium and hence the rate retardation of the reaction is influenced by both temperature and chemical factors. A high temperature favours formation of the fragmentation products rather than the adduct radical P_n-S-C•(Z)-S-P_m. RAFT agents with a radical stabilising Z-group such as Phenyl group favour the adduct radical, as do propagating radicals whose monomers lack radical stabilising features, for example Vinyl acetate.

Further Mechanistic Considerations

In terms of mechanism, an ideal RAFT polymerization has several features. The pre-equilibrium and re-initiation steps are completed very early in the polymerization meaning that the major product of the reaction (the RAFT polymer chains, RAFT-P_n), all start growing at approximately the same time.

The forwards and reverse reactions of the main RAFT equilibrium are fast, favouring equal growth opportunities amongst the chains. The total number of radicals delivered to the system by the initiator during the course of the polymerization is low compared to the number of RAFT agent molecules, meaning that the R group initiated polymer chains from the Re-initiation step form the majority of the chains in the system, rather than initiator fragment bearing chains formed in the Initiation step. This is important because initiator decomposes continuously during the polymerization, not just at the start, and polymer chains arising from initiator decomposition cannot, therefore, have a narrow length distribution. These mechanistic features lead to an average chain length that increases linearly with the conversion of monomer into polymer. In contrast to other controlled radical polymerizations (for example ATRP), A RAFT polymerization does not achieve controlled evolution of molecular weight and low polydispersity by reducing bi-radical termination events (although in some systems, these events may indeed be reduced somewhat, as outlined above), but rather, by ensuring that most polymer chains start growing at approximately the same time and experience equal growth during polymerization.

Block Copolymers

As with other living radical polymerization techniques, RAFT allows chain extension of a polymer of one monomer with a second type of polymer to yield a block copolymer.

In such a polymerisation, there is the additional challenge that the RAFT agent for the first monomer must also be suitable for the second monomer, making block copolymerisation of monomers of highly disparate character challenging.

Multiblock copolymers have also been reported by using difunctional R groups or symmetrical trithiocarbonates with difunctional Z groups.

Smart Materials and Biological Applications

Due to its flexibility with respect to the choice of monomers and reaction conditions, the RAFT process competes favourably with other forms of living polymerization for the generation of bio-materials. New types of polymers are able to be constructed with unique properties, such as temperature and pH sensitivity.

Specific materials and their applications include polymer-protein and polymer-drug conjugates, mediation of enzyme activity, molecular recognition processes and polymeric micelles which can deliver a drug to a specific site in the body.

RAFT has also been used to graft polymer chains onto polymeric surfaces, for example, polymeric microspheres.

RAFT Compared to Other Controlled Polymerizations

Advantages

Polymerization can be performed in large range of solvents (including water), within a wide temperature range, is suitable for use with many different monomers and does not require highly rigorous removal of oxygen and other impurities.

Disadvantages

A particular RAFT agent is only suitable for a limited set of monomers and the synthesis of a RAFT agent typically requires a multistep synthetic procedure and subsequent purification. RAFT agents can be unstable over long time periods, are highly coloured and can have a pungent odour due to gradual decomposition of the dithioester moiety to yield small sulfur compounds. The presence of sulfur and colour in the resulting polymer may also be undesirable for some applications; however, this can, to an extent, be eliminated with further chemical and physical purifiation steps.

Selenium-Centred Radical-Mediated Polymerization

Diphenyl diselenide and several benzylic selenides have been explored by Kwon *et al.* as photoiniferters in polymerization of styrene and methyl methacrylate. Their mechanism of control over polymerization is proposed to be similar to the dithiuram disulfide iniferters. However, their low transfer constants allow them to be used for block copolymer synthesis but give limited control over the molecular weight distribution.

Stibine-Mediated Polymerization

More recently Yamago *et al.* reported stibine-mediated polymerization, using an organostibine transfer agent with the general structure Z(Z')-Sb-R (where Z= activating group and R= free radical leaving group). A wide range of monomers (styrenics, (meth)acrylics and vinylics) can be controlled, giving narrow molecular weight distributions and predictable molecular weights under thermally initiated conditions. Yamago has also published a patent indicating that bismuth alkyls can also control radical polymerizations via a similar mechanism.

Living Group-Transfer Polymerization

Group-transfer polymerization also has characteristics of living polymerization. It is applied to alkylated methacrylate monomers and the initiator is a silyl ketene acetal. New monomer adds to the initiator and to the active growing chain in a Michael reaction.

With each addition of a monomer group the trimethylsilyl group is transferred to the end of the chain. The active chain-end is not ionic as in anionic or cationic polymeriation but is covalent. The reaction can be catalysed by bifluorides and bioxyanions such as *tris(dialkylamino)sulfonium bifluoride* or *tetrabutyl ammonium bibenzoate*. The method was discovered in 1983 by O.W. Webster and the name first suggested by Barry Trost.

Living Ziegler-Natta Polymerization

Several reported methods exist that introduce livingness in Ziegler-Natta polymerization. The monomer in this type of polymerization (a subset of coordination polymerization) is an alpha-olefin and the active site contains an alkyl to metal bond. Chain growth is based on the Cossee-Arlman mechanism.

An early method (Doi, 1979) describes propene polymerization in toluene at -50°C using diethylaluminium chloride and a vanadium catalyst for example $V(acac)_3$ to syndiotactic polypropylene with a polydispersity index of 1.05 to 1.4. Another living system as described by McConville in 1996 is based on titanium using 1-hexene, $[RN(CH_2)_3NR]TiMe_2$ and tris(pentafluorophenyl)boron.

Cobalt-Mediated Radical Polymerization

Cobalt based catalysts, when used in radical polymerization, have several main advantages especially in slowing down the reaction rate, allowing for the synthesis of polymers with peculiar properties. As starting the reaction does need a real radical initiator, the cobalt species is not the only used catalyst, it is a mediator. For this reason this type of reaction is referred to as cobalt mediated radical polymerization. About half of all commercial polymers are produced by radical polymerization. Radical polymerization (RP) reactions have several advantageous properties:

- A wide variety of monomers can be polymerized.
- RP reactions are tolerant to various functional groups.
- RP reactions allow a large temperature range of operation (–100 to >200 °C).
- RP reactions are generally compatibility with several reactions conditions (bulk, solution, (mini)emulsion and suspension).
- RP reactions allow a relatively simple reactor set-up, and are hence costs effective.

However, conventional (free) RP reactions suffer from a lack of control over the polymer molecular-weights and weight distributions. A relatively narrow molecular weight-distribution (M_w/M_w) is usually desirable, as broad distribution negatively influence the polymer properties of (generally atactic) polymers produced by RP. Common RP also does not allow the formation of block copolymers. This is where controlled (or living) radical polymerization comes into play. Several CRP reactions have been developed over the past years, some of which capable of producing well-defined polymers with narrow molecular weight distributions.

Chapter 4

Representation of Structures

Protein structures have to be represented in some coordinate-independent space to make them comparable. This is typically achieved by constructing a sequence-to-sequence matrix or series of matrices that encompass comparative metrics: rather than absolute distances relative to a fixed coordinate space.

An intuitive representation is the distance matrix, which is a two-dimensional matrix containing all pairwise distances between some subset of the atoms in each structure (such as the alpha carbons). The matrix increases in dimensionality as the number of structures to be simultaneously aligned increases.

Reducing the protein to a coarse metric such as secondary structure elements (SSEs) or structural fragments can also produce sensible alignments, despite the loss of information from discarding distances, as noise is also discarded. Choosing a representation to facilitate computation is critical to developing an efficient alignment mechanism. Structural alignment techniques have been used in comparing individual structures or sets of structures as well as in the production of "all-to-all" comparison databases that measure the divergence between every pair of structures present in the Protein Data Bank (PDB). Such databases are used to classify proteins by their fold.

DALI

A common and popular structural alignment method is the DALI, or distance alignment matrix method, which breaks the input structures into hexapeptide fragments and calculates a distance matrix by evaluating the contact patterns between successive fragments.

Secondary structure features that involve residues that are contiguous in sequence appear on the matrix's main diagonal; other diagonals in the matrix reflect spatial contacts between residues that are not near each other in the sequence. When these diagonals are parallel to the main diagonal, the features they represent are parallel; when they are perpendicular, their features are antiparallel. This representation is memory-intensive because the features in the square matrix are symmetrical (and thus redundant) about the main diagonal.

When two proteins' distance matrices share the same or similar features in approximately the same positions, they can be said to have similar folds with similar-length loops connecting their secondary structure elements. DALI's actual alignment process requires a similarity search after the two proteins' distance matrices are built; this is normally conducted via a series of overlapping submatrices of size 6x6. Submatrix matches are then reassembled into a final alignment via a standard score-maximisation algorithm — the original version of DALI used a Monte Carlo simulation to maximise a structural similarity score that is a function of the distances between putative corresponding atoms. In particular, more distant atoms within corresponding features are exponentially downweighted to reduce the effects of noise introduced by loop mobility, helix torsions, and other minor structural variations.

Combinatorial Extension

The combinatorial extension (CE) method is similar to DALI in that it too breaks each structure in the query set into a series of fragments that it then attempts to reassemble into a complete alignment. A series of pairwise combinations of fragments called aligned fragment pairs, or AFPs, are used to define a similarity matrix through which an optimal path is generated to identify the final alignment.

Only AFPs that meet given criteria for local similarity are included in the matrix as a means of reducing the necessary search space and thereby increasing efficiency. A number of similarity metrics are possible; the original definition of the CE method included only structural superpositions and inter-residue distances but has since been expanded to include local environmental properties such as secondary structure, solvent exposure, hydrogen-bonding patterns, and dihedral angles.

An alignment path is calculated as the optimal path through the similarity matrix by linearly progressing through the sequences and extending the alignment with the next possible high-scoring AFP pair. The initial AFP pair that nucleates the alignment can occur at any point in the sequence matrix.

Extensions then proceed with the next AFP that meets given distance criteria restricting the alignment to low gap sizes. The size of each AFP and the maximum gap size are required input parameters but are usually set to empirically determined values of 8 and 30 respectively. Like DALI and SSAP, CE has been used to construct an all-to-all fold classification database from the known protein structures in the PDB.

The RCSB PDB has recently released an updated version of CE and FATCAT as part of the RCSB PDB Protein Comparison Tool. It provides a new variation of CE that can detect circular permutations in protein structures.

GANGSTA+

GANGSTA+ is a combinatorial algorithm for non-sequential structural alignment of proteins and similarity search in databases. It uses a combinatorial approach on the secondary structure level to evaluate similarities between two protein structures based on contact maps. Different SSE assignment modes can be used.

The assignment of SSEs can be performed respecting the sequential order of the SSEs in the polypeptide chains of the considered protein pair (sequential alignment) or by ignoring this order (non-sequential alignment).

Furthermore, SSE pairs can optionally be aligned in reverse orientation. The highest ranking SSE assignments are transferred to the residue level by a pointmatching approach.

To obtain an initial common set of atomic coordinates for both proteins, pairwise attractive interactions of the C-alpha atom pairs are defined by inverse Lorentzians and energy minimised.

MAMMOTH

MAtching Molecular Models Obtained from THeory. As its name suggests, MAMMOTH was originally developed for comparing models coming from structure prediction (THeory) since it is tolerant of large unalignable regions, but it has proven to work well with experimental models, especially when looking for remote homology.

Benchmarks on targets of blind structure prediction (the CASP experiment) and automated GO annotation have shown it is tightly rank correlated with human curated annotation.

A highly complete database of mammoth-based structure annotations for the predicted structures of unknown proteins covering 150 genomes facilitates genomic scale normalization.

MAMMOTH-based structure alignment methods decompose the protein structure into short, seven-residue, peptides (heptapeptides) which are compared with the heptapeptides of another protein. The similarity score between two heptapeptides is calculated using a unit-vector RMS (URMS) method.

These scores are stored in a similarity matrix, and with a hybrid (local-global) dynamic programming the optimal residue alignment is calculated. Protein similarity scores calculated with MAMMOTH is derived from the likelihood of obtaining a given structural alignment by chance.

MAMMOTH-mult is an extension of the MAMMOTH algorithm to be used to align related families of protein structures. This algorithm is very fast and produces consistent and high quality structural alignments. Multiple structural alignments calculated with MAMMOTH-mult produces structurally implied sequence alignments that can be further used for multiple-template homology modelling, HMM-based protein structure prediction, and profile-type PSI-BLAST searches.

ProBiS

Protein Binding Sites. ProBiS detects structurally similar sites on protein surfaces by local surface structure alignment. It compares the query protein to members of a database of protein 3D structures and detects with sub-residue precision, structurally similar sites as patterns of physicochemical properties on the protein surface. Using an efficient maximum clique algorithm, the program identifies proteins that share local structural similarities with the query protein and generates structure-based alignments of these proteins with the query. Structural similarity scores are calculated for the query protein's surface residues, and are expressed as different colours on the query protein surface. The algorithm has been used successfully for the detection of protein–protein, protein–small ligand and protein–DNA binding sites.

RAPIDO

Rapid Alignment of Proteins In terms of DOmains. RAPIDO is a web server for the 3D alignment of crystal structures of different protein molecules, in the presence of conformational changes. Similar to what CE does as a first step, RAPIDO identifies fragments that are structurally similar in the two proteins using an approach based on difference distance matrices.

The Matching Fragment Pairs (MFPs) are then represented as nodes in a graph which are chained together to form an alignment by means of an algorithm for the identification of the longest path on a DAG (Directed Acyclic Graph). The final step of refinement is performed to improve the quality of the alignment.

After aligning the two structures the server applies a genetic algorithm for the identification of conformationally invariant regions. These regions correspond to groups of atoms whose interatomic distances are constant (within a defined tolerance).

In doing so RAPIDO takes into account the variation in the reliability of atomic coordinates by employing weighting-functions based on the refined B-values. The regions identified as conformationally invariant by RAPIDO represent reliable sets of atoms for the superposition of the two structures that can be used for a detailed analysis of changes in the conformation. In addition to the functionalities provided by existing tools, RAPIDO can identify structurally equivalent regions even when these consist of fragments that are distant in terms of sequence and separated by other movable domains.

SABERTOOTH

SABERTOOTH uses structural profiles to perform structural alignments. The underlying structural profiles expresses the global connectivity of each residue. Despite the very condensed vectorial representation, the tool recognises structural similarities with accuracy comparable to established alignment tools based on coordinates and performs comparably in quality.

Furthermore, the algorithm has favourable scaling of computation time with chain length. Since the algorithm is independent of the details of the structural representation, thc framework can be generalised to sequence-to-sequence and sequence-to-structure comparison within the same setup, and it is therefore more general than other tools.

SSAP

The SSAP (Sequential Structure Alignment Program) method uses double dynamic programming to produce a structural alignment based on atom-to-atom vectors in structure space. Instead of the alpha carbons typically used in structural alignment, SSAP constructs its vectors from the beta carbons for all residues except glycine, a method which thus takes into account the rotameric state of each residue as well as its location along the backbone. SSAP works by first constructing a series of inter-residue distance vectors between each residue and its nearest non-contiguous neighbours on each protein.

A series of matrices are then constructed containing the vector differences between neighbours for each pair of residues for which vectors were constructed. Dynamic programming applied to each resulting matrix determines a series of optimal local alignments which are then summed into a "summary" matrix to which dynamic programming is applied again to determine the overall structural alignment.

SSAP originally produced only pairwise alignments but has since been extended to multiple alignments as well. It has been applied in an all-to-all fashion to produce a hierarchical fold classification scheme known as CATH (Class, Architecture, Topology, Homology), which has been used to construct the CATH Protein Structure Classification database.

SPalign

SPalign is based on a new size-independent score called SPscore for pairwise protein structure alignment. SP-score fixes the cutoff distance at 4 Å and removed the size dependence using a normalization pre-factor. The method is compared to DALI, CE, TMalign and others in fold recognition and function prediction ().

TOPOFIT

In the TOPOFIT method, of protein structures is analysed using three-dimensional Delaunay triangulation patterns derived from backbone representation. It has been found that structurally related proteins have a common spatial invariant part, a set of tetrahedrons, mathematically described as a common spatial sub-graph volume of the three-dimensional contact graph derived from Delaunay tessellation (DT).

Based on this property of protein structures we present a novel common volume superimposition (TOPOFIT) method to produce structural alignments of proteins. The superimposition of the DT patterns allows one to objectively identify a common number of equivalent residues in the structural alignment, in other words, TOPOFIT identifies a feature point on the RMSD/Ne curve, a topomax point, until which two structures correspond to each other including backbone and inter-residue contacts, while the growing number of mismatches between the DT patterns occurs at larger RMSD (Ne) after topomax point.

The topomax point is present in all alignments from different protein structural classes; therefore, the TOPOFIT method identifies common, invariant structural parts between proteins.

The TOPOFIT method adds new opportunities for the comparative analysis of protein structures and for more detailed studies on understanding the molecular principles of tertiary structure organisation and functionality.

It helps to detect conformational changes, topological differences in variable parts, which are particularly important for studies of variations in active/binding sites and protein classification.

SSM

Secondary Structure Matching (SSM), or PDBeFold at the Protein Data Bank in Europe uses graph matching followed by c-alpha alignment to compute alignments.

Recent Developments

Improvements in structural alignment methods constitute an active area of research, and new or modified methods are often proposed that are claimed to offer advantages over the older and more widely distributed techniques. A recent example, TM-align, uses a novel method for weighting its distance matrix, to which standard dynamic programming is then applied.

The weighting is proposed to accelerate the convergence of dynamic programming and correct for effects arising from alignment lengths. In a benchmarking study, TM-align has been reported to improve in both speed and accuracy over DALI and CE.

TM-score in TM-align, however, implicitly assumes that two structures align in a length of L (shorter length or average length). In reality, it is quite possible that only a part of a protein of unknown size aligns with a part of another protein, particularly for multidomain proteins. SP-score is proposed to remove the size dependence of alignment score not by size-dependent cutoff but by a size-dependent normalization factor.

This allows introduction of an effective alignment length that removes the need to specify a length for normalization. For fold discrimination, SPalign improves sensitivity over TM-align for the chain-level comparison by 12% and over DALI for the domain-level comparison by 13% at the same specificity of 99.6%.

However, as algorithmic improvements and computer performance have erased purely technical deficiencies in older approaches, it has become clear that there is no one universal criterion for the 'optimal' structural alignment. TM-align, for instance, is particularly robust in quantifying comparisons between sets of proteins with great disparities in sequence lengths, but it only indirectly captures hydrogen bonding or secondary structure order conservation which might be better metrics for alignment of evolutionarily related proteins.

Thus recent developments have focused on optimizing particular attributes such as speed, quantification of scores, correlation to alternative gold standards, or tolerance of imperfection in structural data or ab initio structural models. An alternative methodology that is gaining popularity is to use the *consensus* of various methods to ascertain proteins structural similarities.

In choosing among modern algorithms, investigators should strongly consider the optimization for the purpose of intended application, as well as on the algorithm's penetration into the particular field so as to facilitate comparison to other authors' results.

For example, MAMMOTH has been specialised for speed and correlation to human annotation, and is thus suited for large-scale structural genomics studies. The venerable DALI is perhaps the most ubiquitous in the literature and due to its integration with other European Bioinformatics Institute web-based tools, the EBI DALI is easily approached by researchers interested in singleton applications.

Historically, it was initially unclear if comparisons that preserved sequence order would be more sensitive than ones that simply compare architecture or contacts without regard to secondary segment ordering. Early versions of DALI allowed a choice of non-sequential alignment at a great cost in speed. Non-sequential methods lost favour as segment order preserving methods outperformed them in speed, quantification of high-confidence similarity scores, and amenability to adoption of rich scoring heuristics. However, in some applications, discovery of conserved but out-of-order structure motif recognition is vital and additionally some forms of experimental data collection, such as cryo-electron microscopy, generally resolve regular secondary elements but not their connection order. This has renewed interest in Non-sequential approaches. Some examples are GANGSTA+ and TOPOFIT.

RNA Structural Alignment

Structural alignment techniques have traditionally been applied exclusively to proteins, as the primary biological macromolecules that assume characteristic three-dimensional structures. However, large RNA molecules also form characteristic tertiary structures, which are mediated primarily by hydrogen bonds formed between base pairs as well as base stacking. Functionally similar noncoding RNA molecules can be especially difficult to extract from genomics data because structure is more strongly conserved than sequence in RNA as well as in proteins, and the more limited alphabet of RNA decreases the information content of any given nucleotide at any given position. A recent method for pairwise structural alignment of RNA sequences with low sequence identity has been published and implemented in the program FOLDALIGN. However, this method is not truly analogous to protein structural alignment techniques because it computationally predicts the structures of the RNA input sequences rather than requiring experimentally determined structures as input. Although computational prediction of the protein folding process has not been particularly successful to date, RNA structures without pseudoknots can often be sensibly predicted using free energy-based scoring methods that account for base pairing and stacking.

DALI

The DALI method, or distance matrix alignment, is a fragment-based method for constructing structural alignments based on contact similarity patterns between successive hexapeptides in the query sequences. It can generate pairwise or multiple alignments and identify a query sequence's structural neighbours in the Protein Data Bank (PDB). It has been used to construct the FSSP structural alignment database (Fold classification based on Structure-Structure alignment of Proteins, or Families of Structurally Similar Proteins). A DALI webserver can be accessed at EBI DALI and the FSSP is located at The Dali Database.

SSAP

SSAP (sequential structure alignment program) is a dynamic programming-based method of structural alignment that uses atom-to-atom vectors in structure space as comparison points. It has been extended since its original description to include multiple as well as pairwise alignments, and has been used in the construction of the CATH (Class, Architecture, Topology, Homology) hierarchical database classification of protein folds. The CATH database can be accessed at CATH Protein Structure Classification.

Combinatorial Extension

The combinatorial extension method of structural alignment generates a pairwise structural alignment by using local geometry to align short fragments of the two proteins being analysed and then assembles these fragments into a larger alignment. Based on measures such as rigid-body root mean square distance, residue distances, local secondary structure, and surrounding environmental features such as residue neighbour hydrophobicity, local alignments called "aligned fragment pairs" are generated and used to build a similarity matrix representing all possible structural alignments within predefined cutoff criteria. A path from one protein structure state to the other is then traced through the matrix by extending the growing alignment one fragment at a time. The optimal such path defines the combinatorial-extension alignment. A web-based server implementing the method and providing a database of pairwise alignments of structures in the Protein Data Bank is located at the Combinatorial Extension website.

Phylogenetic Analysis

Phylogenetics and sequence alignment are closely related fields due to the shared necessity of evaluating sequence relatedness. The field of phylogenetics makes extensive use of sequence alignments in the construction and interpretation of phylogenetic trees, which are used to classify the evolutionary relationships between homologous genes represented in the genomes of divergent species. The degree to which sequences in a query set differ is qualitatively related to the sequences' evolutionary distance from one another. Roughly speaking, high sequence identity suggests that the sequences in question have a comparatively young most recent common ancestor, while low identity suggests that the divergence is more ancient. This approximation, which reflects the "molecular clock" hypothesis that a roughly constant rate of evolutionary change can be used to extrapolate the elapsed time since two genes first diverged (that is, the coalescence time), assumes that the effects of mutation and selection are constant across sequence lineages. Therefore it does not account for possible difference among organisms or species in the rates of DNA repair or the possible functional conservation of specific regions in a sequence.

(In the case of nucleotide sequences, the molecular clock hypothesis in its most basic form also discounts the difference in acceptance rates between silent mutations that do not alter the meaning of a given codon and other mutations that result in a different amino acid being incorporated into the protein.) More statistically accurate methods allow the evolutionary rate on each branch of the phylogenetic tree to vary, thus producing better estimates of coalescence times for genes. Progressive multiple alignment techniques produce a phylogenetic tree by necessity because they incorporate sequences into the growing alignment in order of relatedness. Other techniques that assemble multiple sequence alignments and phylogenetic trees score and sort trees first and calculate a multiple sequence alignment from the highest-scoring tree. Commonly used methods of phylogenetic tree construction are mainly heuristic because the problem of selecting the optimal tree, like the problem of selecting the optimal multiple sequence alignment, is NP-hard.

Computational phylogenetics is the application of computational algorithms, methods and programs to phylogenetic analyses. The goal is to assemble a phylogenetic tree representing a hypothesis about the evolutionary ancestry of a set of genes, species, or other taxa.

For example, these techniques have been used to explore the family tree of hominid species and the relationships between specific genes shared by many types of organisms. Traditional phylogenetics relies on morphological data obtained by measuring and quantifying the phenotypic properties of representative organisms, while the more recent field of molecular phylogenetics uses nucleotide sequences encoding genes or amino acid sequences encoding proteins as the basis for classification.

Many forms of molecular phylogenetics are closely related to and make extensive use of sequence alignment in constructing and refining phylogenetic trees, which are used to classify the evolutionary relationships between homologous genes represented in the genomes of divergent species.

The phylogenetic trees constructed by computational methods are unlikely to perfectly reproduce the evolutionary tree that represents the historical relationships between the species being analysed. The historical species tree may also differ from the historical tree of an individual homologous gene shared by those species.

Producing a phylogenetic tree requires a measure of homology among the characteristics shared by the taxa being compared. In morphological studies, this requires explicit decisions about which physical characteristics to measure and how to use them to encode distinct states corresponding to the input taxa. In molecular studies, a primary problem is in producing a multiple sequence alignment (MSA) between the genes or amino acid sequences of interest. Progressive sequence alignment methods produce a phylogenetic tree by necessity because they incorporate new sequences into the calculated alignment in order of genetic distance.

Types of Phylogenetic Trees

Phylogenetic trees generated by computational phylogenetics can be either *rooted* or *unrooted* depending on the input data and the algorithm used. A rooted tree is a directed graph that explicitly identifies a most recent common ancestor (MRCA), usually an imputed sequence that is not represented in the input. Genetic distance measures can be used to plot a tree with the input sequences as leaf nodes and their distances from the root proportional to their genetic distance from the hypothesized MRCA. Identification of a root usually requires the inclusion in the input data of at least one "outgroup" known to be only distantly related to the sequences of interest.

By contrast, unrooted trees plot the distances and relationships between input sequences without making assumptions regarding their descent. An unrooted tree can always be produced from a rooted tree, but a root cannot usually be placed on an unrooted tree without additional data on divergence rates, such as the assumption of the molecular clock hypothesis. The set of all possible phylogenetic trees for a given group of input sequences can be conceptualized as a discretely defined multidimensional "tree space" through which search paths can be traced by optimization algorithms. Although counting the total number of trees for a nontrivial number of input sequences can be complicated by variations in the definition of a tree topology, it is always true that there are more rooted than unrooted trees for a given number of inputs and choice of parameters.

Coding Characters and Defining Homology

Morphological Analysis

The basic problem in morphological phylogenetics is the assembly of a matrix representing a mapping from each of the taxa being compared to representative measurements for each of the phenotypic characteristics being used as a classifier.

The types of phenotypic data used to construct this matrix depend on the taxa being compared; for individual species, they may involve measurements of average body size, lengths or sizes of particular bones or other physical features, or even behavioural manifestations. Of course, since not every possible phenotypic characteristic could be measured and encoded for analysis, the selection of which features to measure is a major inherent obstacle to the method.

The decision of which traits to use as a basis for the matrix necessarily represents a hypothesis about which traits of a species or higher taxon are evolutionarily relevant. Morphological studies can be confounded by examples of convergent evolution of phenotypes. A major challenge in constructing useful classes is the high likelihood of inter-taxon overlap in the distribution of the phenotype's variation.

The inclusion of extinct taxa in morphological analysis is often difficult due to absence of or incomplete fossil records, but has been shown to have a significant effect on the trees produced; in one study only the inclusion of extinct species of apes produced a morphologically derived tree that was consistent with that produced from molecular data.

Some phenotypic classifications, particularly those used when analysing very diverse groups of taxa, are discrete and unambiguous; classifying organisms as possessing or lacking a tail, for example, is straightforward in the majority of cases, as is counting features such as eyes or vertebrae. However, the most appropriate representation of continuously varying phenotypic measurements is a controversial problem without a general solution. A common method is simply to sort the measurements of interest into two or more classes, rendering continuous observed variation as discretely classifiable (e.g., all examples with humerus bones longer than a given cutoff are scored as members of one state, and all members whose humerus bones are shorter than the cutoff are scored as members of a second state). This results in an easily manipulated data set but has been criticized for poor reporting of the basis for the class definitions and for sacrificing information compared to methods that use a continuous weighted distribution of measurements. Because morphological data is extremely labour-intensive to collect, whether from literature sources or from field observations, reuse of previously compiled data matrices is not uncommon, although this may propagate flaws in the original matrix into multiple derivative analyses.

Molecular Analysis

The problem of character coding is very different in molecular analyses, as the characters in biological sequence data are immediate and discretely defined - distinct nucleotides in DNA or RNA sequences and distinct amino acids in protein sequences. However, defining homology can be challenging due to the inherent difficulties of multiple sequence alignment. For a given gapped MSA, several rooted phylogenetic trees can be constructed that vary in their interpretations of which changes are "mutations" versus ancestral characters, and which events are insertion mutations or deletion mutations. For example, given only a pairwise alignment with a gap region, it is impossible to determine whether one sequence bears an insertion mutation or the other carries a deletion. The problem is magnified in MSAs with unaligned and nonoverlapping gaps. In practice, sizable regions of a calculated alignment may be discounted in phylogenetic tree construction to avoid integrating noisy data into the tree calculation.

Distance-Matrix Methods

Distance-matrix methods of phylogenetic analysis explicitly rely on a measure of "genetic distance" between the sequences being classified, and therefore they require an MSA as an input. Distance is often defined as the fraction of mismatches at aligned positions, with gaps either ignored or counted as mismatches. Distance methods attempt to construct an all-to-all matrix from the sequence query set describing the distance between each sequence pair. From this is constructed a phylogenetic tree that places closely related sequences under the same interior node and whose branch lengths closely reproduce the observed distances between sequences. Distance-matrix methods may produce either rooted or unrooted trees, depending on the algorithm used to calculate them. They are frequently used as the basis for progressive and iterative types of multiple sequence alignments. The main disadvantage of distance-matrix methods is their inability to efficiently use information about local high-variation regions that appear across multiple subtrees.

Neighbour-Joining

Neighbour-joining methods apply general data clustering techniques to sequence analysis using genetic distance as a clustering metric. The simple neighbour-joining method produces unrooted trees, but it does not assume a constant rate of evolution (i.e., a molecular clock) across lineages. Its relative, UPGMA (Unweighted Pair Group Method with Arithmetic mean) produces rooted trees and requires a constant-rate assumption - that is, it assumes an ultrametric tree in which the distances from the root to every branch tip are equal.

Fitch-Margoliash Method

The Fitch-Margoliash method uses a weighted least squares method for clustering based on genetic distance. Closely related sequences are given more weight in the tree construction process to correct for the increased inaccuracy in measuring distances between distantly related sequences. The distances used as input to the algorithm must be normalized to prevent large artifacts in computing relationships between closely related and distantly related groups. The distances calculated by this method must be linear; the linearity criterion for distances requires that the expected values of the branch lengths for two individual branches must equal the expected value of the sum of the two branch distances - a property that applies to biological sequences only when they have been corrected for the possibility of back mutations at individual sites. This correction is done through the use of a substitution matrix such as that derived from the Jukes-Cantor model of DNA evolution. The distance correction is only necessary in practice when the evolution rates differ among branches.

Another modification of the algorithm can be helpful, especially in case of concentrated distances (please report to Concentration of measure phenomenon and Curse of dimensionality): that modification, described in , has been shown to improve the efficiency of the algorithm and its robustness.

The least-squares criterion applied to these distances is more accurate but less efficient than the neighbour-joining methods. An additional improvement that corrects for correlations between distances that arise from many closely related sequences in the data set can also be applied at increased computational cost. Finding the optimal least-squares tree with any correction factor is NP-complete, so heuristic search methods like those used in maximum-parsimony analysis are applied to the search through tree space.

Using Outgroups

Independent information about the relationship between sequences or groups can be used to help reduce the tree search space and root unrooted trees. Standard usage of distance-matrix methods involves the inclusion of at least one outgroup sequence known to be only distantly related to the sequences of interest in the query set.

This usage can be seen as a type of experimental control. If the outgroup has been appropriately chosen, it will have a much greater genetic distance and thus a longer branch length than any other sequence, and it will appear near the root of a rooted tree. Choosing an appropriate outgroup requires the selection of a sequence that is moderately related to the sequences of interest; too close a relationship defeats the purpose of the outgroup and too distant adds noise to the analysis. Care should also be taken to avoid situations in which the species from which the sequences were taken are distantly related, but the gene encoded by the sequences is highly conserved across lineages. Horizontal gene transfer, especially between otherwise divergent bacteria, can also confound outgroup usage.

Maximum Parsimony

Maximum parsimony (MP) is a method of identifying the potential phylogenetic tree that requires the smallest total number of evolutionary events to explain the observed sequence data. Some ways of scoring trees also include a "cost" associated with particular types of evolutionary events and attempt to locate the tree with the smallest total cost. This is a useful approach in cases where not every possible type of event is equally likely - for example, when particular nucleotides or amino acids are known to be more mutable than others.

The most naive way of idcntifying the most parsimonious tree is simple enumeration - considering each possible tree in succession and searching for the tree with the smallest score. However, this is only possible for a relatively small number of sequences or species because the problem of identifying the most parsimonious tree is known to be NP-hard; consequently a number of heuristic search methods for optimization have been developed to locate a highly parsimonious tree, if not the best in the set. Most such methods involve a steepest descent-style minimisation mechanism operating on a tree rearrangement criterion.

Branch and Bound

The branch and bound algorithm is a general method used to increase the efficiency of searches for near-optimal solutions of NP-hard problems first applied to phylogenetics in the early 1980s. Branch and bound is particularly well suited to phylogenetic tree construction because it inherently requires dividing a problem into a tree structure as it subdivides the problem space into smaller regions.

As its name implies, it requires as input both a branching rule (in the case of phylogenetics, the addition of the next species or sequence to the tree) and a bound (a rule that excludes certain regions of the search space from consideration, thereby assuming that the optimal solution cannot occupy that region). Identifying a good bound is the most challenging aspect of the algorithm's application to phylogenetics. A simple way of defining the bound is a maximum number of assumed evolutionary changes allowed per tree. A set of criteria known as Zharkikh's rules severely limit the search space by defining characteristics shared by all candidate "most parsimonious" trees. The two most basic rules require the elimination of all but one redundant sequence (for cases where multiple observations have produced identical data) and the elimination of character sites at which two or more states do not occur in at least two species. Under ideal conditions these rules and their associated algorithm would completely define a tree.

Sankoff-Morel-Cedergren Algorithm

The Sankoff-Morel-Cedergren algorithm was among the first published methods to simultaneously produce an MSA and a phylogenetic tree for nucleotide sequences.

The method uses a maximum parsimony calculation in conjunction with a scoring function that penalizes gaps and mismatches, thereby favouring the tree that introduces a minimal number of such events. The imputed sequences at the interior nodes of the tree are scored and summed over all the nodes in each possible tree. The lowest-scoring tree sum provides both an optimal tree and an optimal MSA given the scoring function. Because the method is highly computationally intensive, an approximate method in which initial guesses for the interior alignments are refined one node at a time. Both the full and the approximate version are in practice calculated by dynamic programming.

MALIGN and POY

More recent phylogenetic tree/MSA methods use heuristics to isolate high-scoring, but not necessarily optimal, trees. The MALIGN method uses a maximum-parsimony technique to compute a multiple alignment by maximising a cladogram score, and its companion POY uses an iterative method that couples the optimization of the phylogenetic tree with improvements in the corresponding MSA. However, the use of these methods in constructing evolutionary hypotheses has been criticized as biased due to the deliberate construction of trees reflecting minimal evolutionary events.

Maximum Likelihood

The maximum likelihood method uses standard statistical techniques for inferring probability distributions to assign probabilities to particular possible phylogenetic trees. The method requires a substitution model to assess the probability of particular mutations; roughly, a tree that requires more mutations at interior nodes to explain the observed phylogeny will be assessed as having a lower probability.

This is broadly similar to the maximum-parsimony method, but maximum likelihood allows additional statistical flexibility by permitting varying rates of evolution across both lineages and sites.

In fact, the method requires that evolution at different sites and along different lineages must be statistically independent. Maximum likelihood is thus well suited to the analysis of distantly related sequences, but because it formally requires search of all possible combinations of tree topology and branch length, it is computationally expensive to perform on more than a few sequences.

The "pruning" algorithm, a variant of dynamic programming, is often used to reduce the search space by efficiently calculating the likelihood of subtrees. The method calculates the likelihood for each site in a "linear" manner, starting at a node whose only descendants are leaves (that is, the tips of the tree) and working backwards toward the "bottom" node in nested sets. However, the trees produced by the method are only rooted if the substitution model is irreversible, which is not generally true of biological systems. The search for the maximum-likelihood tree also includes a branch length optimization component that is difficult to improve upon algorithmically; general global optimization tools such as the Newton-Raphson method are often used. Searching tree topologies defined by likelihood has not been shown to be NP-complete, but remains extremely challenging because branch-and-bound search is not yet effective for trees represented in this way.

Bayesian Inference

Bayesian inference can be used to produce phylogenetic trees in a manner closely related to the maximum likelihood methods. Bayesian methods assume a prior probability distribution of the possible trees, which may simply be the probability of any one tree among all the possible trees that could be generated from the data, or may be a more sophisticated estimate derived from the assumption that divergence events such as speciation occur as stochastic processes.

The choice of prior distribution is a point of contention among users of Bayesian-inference phylogenetics methods. Implementations of Bayesian methods generally use Markov chain Monte Carlo sampling algorithms, although the choice of move set varies; selections used in Bayesian phylogenetics include circularly permuting leaf nodes of a proposed tree at each step and swapping descendant subtrees of a random internal node between two related trees.

The use of Bayesian methods in phylogenetics has been controversial, largely due to incomplete specification of the choice of move set, acceptance criterion, and prior distribution in published work.

Model Selection

Molecular phylogenetics methods rely on a defined substitution model that encodes a hypothesis about the relative rates of mutation at various sites along the gene or amino acid sequences being studied. At their simplest, substitution models aim to correct for differences in the rates of transitions and transversions in nucleotide sequences.

The use of substitution models is necessitated by the fact that the genetic distance between two sequences increases linearly only for a short time after the two sequences diverge from each other (alternatively, the distance is linear only shortly before coalescence).

The longer the amount of time after divergence, the more likely it becomes that two mutations occur at the same nucleotide site. Simple genetic distance calculations will thus undercount the number of mutation events that have occurred in evolutionary history.

The extent of this undercount increases with increasing time since divergence, which can lead to the phenomenon of long branch attraction, or the misassignment of two distantly related but convergently evolving sequences as closely related.

The maximum parsimony method is particularly susceptible to this problem due to its explicit search for a tree representing a minimum number of distinct evolutionary events.

Types of Models

All substitution models assign a set of weights to each possible change of state represented in the sequence. The most common model types are implicitly reversible because they assign the same weight to, for example, a G>C nucleotide mutation as to a C>G mutation. The simplest possible model, the Jukes-Cantor model, assigns an equal probability to every possible change of state for a given nucleotide base.

The rate of change between any two distinct nucleotides will be one-third of the overall substitution rate. More advanced models distinguish between transitions and transversions. The most general possible time-reversible model, called the GTR model, has six mutation rate parameters.

An even more generalised model known as the general 12-parameter model breaks time-reversibility, at the cost of much additional complexity in calculating genetic distances that are consistent among multiple lineages. One possible variation on this theme adjusts the rates so that overall GC content - an important measure of DNA double helix stability - varies over time.

Models may also allow for the variation of rates with positions in the input sequence. The most obvious example of such variation follows from the arrangement of nucleotides in protein-coding genes into three-base codons.

If the location of the open reading frame (ORF) is known, rates of mutation can be adjusted for position of a given site within a codon, since it is known that wobble base pairing can allow for higher mutation rates in the third nucleotide of a given codon without affecting the codon's meaning in the genetic code.

A less hypothesis-driven example that does not rely on ORF identification simply assigns to each site a rate randomly drawn from a predetermined distribution, often the gamma distribution or log-normal distribution. Finally, a more conservative estimate of rate variations known as the covarion method allows autocorrelated variations in rates, so that the mutation rate of a given site is correlated across sites and lineages. In biology, a substitution model describes the process from which a sequence of characters changes into another set of traits. For example, in cladistics, each position in the sequence might correspond to a property of a species which can either be present or absent. The alphabet could then consist of "0" for absence and "1" for presence. Then the sequence 00110 could mean, for example, that a species does not have feathers or lay eggs, does have fur, is warm-blooded, and cannot breathe underwater. Another sequence 11010 would mean that a species has feathers, lays eggs, does not have fur, is warm-blooded, and cannot breathe underwater. In phylogenetics, sequences are often obtained by firstly obtaining a nucleotide or protein sequence alignment, and then taking the bases or amino acids at corresponding positions in the alignment as the characters. Sequences achieved by this might look like AGCGGAGCTTA and GCCGTAGACGC.

Substitution models are used for a number of things:

1. Constructing evolutionary trees in phylogenetics or cladistics.
2. Simulating sequences to test other methods and algorithms.

The Molecular Clock and the Units of Time

Typically, a branch length of a phylogenetic tree is expressed as the expected number of substitutions per site; if the evolutionary model indicates that each site within an ancestral sequence will typically experience x substitutions by the time it evolves to a particular descendant's sequence then the ancestor and descendant are considered to be separated by branch length x. Sometimes a branch length is measured in terms of geological years. For example, a fossil record may make it possible to determine the number of years between an ancestral species and a descendant species. Because some species evolve at faster rates than others, these two measures of branch length are not always in direct proportion. The expected number of substitutions per site per year is often indicated with the Greek letter mu (μ). A model is said to have a strict molecular clock if the expected number of substitutions per year μ is constant regardless of which species' evolution is being examined. An important implication of a strict molecular clock is that the number of expected substitutions between an ancestral species and any of its present-day descendants must be independent of which descendant species is examined. Note that the assumption of a strict molecular clock is often unrealistic, especially across long periods of evolution. For example, even though rodents are genetically very similar to primates, they have undergone a much higher number of substitutions in the estimated time since divergence in some regions of the genome. This could be due to their shorter generation time, higher metabolic rate, increased population structuring, increased rate of speciation, or smaller body size. When studying ancient events like the Cambrian explosion under a molecular clock assumption, poor concurrence between cladistic and phylogenetic data is often observed. There has been some work on models allowing variable rate of evolution.

Models that can take into account variability of the rate of the molecular clock between different evolutionary lineages in the phylogeny are called "relaxed" in opposition to "strict". In such models the rate can be assumed to be correlated or not between ancestors and descendants and rate variation among lineages can be drawn from many distributions but usually exponential and lognormal distributions are applied. There is a special case, called "local molecular clock" when a phylogeny is divided into at least two partitions (sets of lineages) and in each strict molecular clock is applied but with different rate.

Mechanistic vs. Empirical Models

A main difference in evolutionary models is how many parameters are estimated every time for the data set under consideration and how many of them are estimated once on a large data set. Mechanistic models describe all substitutions as a function of a number of parameters which are estimated for every data set analysed, preferably using maximum likelihood.

This has the advantage that the model can be adjusted to the particularities of a specific data set (e.g. different composition biases in DNA). Problems can arise when too many parameters are used, particularly if they can compensate for each other. Then it is often the case that the data set is too small to yield enough information to estimate all parameters accurately.

Empirical models are created by estimating many parameters (typically all entries of the rate matrix and the character frequencies) from a large data set. These parameters are then fixed and will be reused for every data set. This has the advantage that those parameters can be estimated more accurately. Normally, it is not possible to estimate all entries of the substitution matrix from the current data set only. On the downside, the estimated parameters might be too generic and do not fit a particular data set well enough. With the large-scale genome sequencing still producing very large amounts of DNA and protein sequences, there is enough data available to create empirical models with any number of parameters. Because of the problems mentioned above, the two approaches are often combined, by estimating most of the parameters once on large-scale data, while a few remaining parameters are then adjusted to the data set under consideration. The following sections give an overview of the different approaches taken for DNA, protein or codon-based models.

Models of DNA Substitution

Models of DNA evolution were first proposed in 1969 by Jukes and Cantor, assuming equal transition rates as well as equal equilibrium frequencies for all bases. In 1980 Kimura introduced a model with two parameters: one for the transition and one for the transversion rate and in 1981, Felsenstein made a model in which the substitution rate corresponds to the equilibrium frequency of the target nucleotide. Hasegawa, Kishino and Yano (HKY) unified the two last models to a six parameter model. In the 1990s, models similar to HKY were developed and refined by several researchers. For DNA substitution models, mainly mechanistic models (as described above) are employed. The small number of parameters to estimate makes this feasible, but also DNA is often highly optimized for specific purposes (e.g. fast expression or stability) depending on the organism and the type of gene, making it necessary to adjust the model to these circumstances.

Models of Amino Acid Substitutions

For many analyses, particularly for longer evolutionary distances, the evolution is modelled on the amino acid level. Since not all DNA substitution also alter the encoded amino acid, information is lost when looking at amino acids instead of nucleotide bases. However, several advantages speak in favour of using the amino acid information: DNA is much more inclined to show compositional bias than amino acids, not all positions in the DNA evolve at the same speed (non-synonymous mutations are more likely to become fixed in the population than synonymous ones), but probably most important, because of those fast evolving positions and the limited alphabet size (only four possible states), the DNA suffers much more from back substitutions, making it difficult to accurately estimate longer distances.

Unlike the DNA models, amino acid models traditionally are empirical models. They were pioneered in the 1970s by Dayhoff and co-workers , by estimating replacement rates from protein alignments with at least 85% identity.

This minimised the chances of observing multiple substitutions at a site. From the estimated rate matrix, a series of replacement probability matrices were derived, known under names such as PAM250. The Dayhoff model was used to assess the significance of homology search results, but also for phylogenetic analyses. The Dayhoff PAM matrices were based on relatively few alignments (since not more were available at that time), but in the 1990s, new matrices were estimated using almost the same methodology, but based on the large protein databases available then (, the latter being known as "JTT" matrices).

Choosing the Best Model

The selection of an appropriate model is critical for the production of good phylogenetic analyses, both because underparameterized or overly restrictive models may produce aberrant behaviour when their underlying assumptions are violated, and because overly complex or overparameterized models are computationally expensive and the parameters may be overfit. The most common method of model selection is the likelihood ratio test (LRT), which produces a likelihood estimate that can be interpreted as a measure of "goodness of fit" between the model and the input data.

However, care must be taken in using these results, since a more complex model with more parameters will always have a higher likelihood than a simplified version of the same model, which can lead to the naive selection of models that are overly complex.

For this reason model selection computer programs will choose the simplest model that is not significantly worse than more complex substitution models. A significant disadvantage of the LRT is the necessity of making a series of pairwise comparisons between models; it has been shown that the order in which the models are compared has a major effect on the one that is eventually selected.

An alternative model selection method is the Akaike information criterion (AIC), formally an estimate of the Kullback-Leibler divergence between the true model and the model being tested. It can be interpreted as a likelihood estimate with a correction factor to penalize overparameterized models. The AIC is calculated on an individual model rather than a pair, so it is independent of the order in which models are assessed. A related alternative, the Bayesian information criterion (BIC), has a similar basic interpretation but penalizes complex models more heavily.

Assessment of Significance

Sequence alignments are useful in bioinformatics for identifying sequence similarity, producing phylogenetic trees, and developing homology models of protein structures. However, the biological relevance of sequence alignments is not always clear.

Alignments are often assumed to reflect a degree of evolutionary change between sequences descended from a common ancestor; however, it is formally possible that convergent evolution can occur to produce apparent similarity between proteins that are evolutionarily unrelated but perform similar functions and have similar structures.

In database searches such as BLAST, statistical methods can determine the likelihood of a particular alignment between sequences or sequence regions arising by chance given the size and composition of the database being searched. These values can vary significantly depending on the search space. In particular, the likelihood of finding a given alignment by chance increases if the database consists only of sequences from the same organism as the query sequence. Repetitive sequences in the database or query can also distort both the search results and the assessment of statistical significance; BLAST automatically filters such repetitive sequences in the query to avoid apparent hits that are statistical artifacts.

Methods of statistical significance estimation for gapped sequence alignments are available in the literature.

Assessment of Credibility

Statistical significance indicates the probability that an alignment of a given quality could arise by chance, but does not indicate how much superior a given alignment is to alternative alignments of the same sequences. Measures of alignment credibility indicate the extent to which the best scoring alignments for a given pair of sequences are substantially similar. Methods of alignment credibility estimation for gapped sequence alignments are available in the literature.

Scoring Functions

The choice of a scoring function that reflects biological or statistical observations about known sequences is important to producing good alignments. Protein sequences are frequently aligned using substitution matrices that reflect the probabilities of given character-to-character substitutions.

A series of matrices called PAM matrices (Point Accepted Mutation matrices, originally defined by Margaret Dayhoff and sometimes referred to as "Dayhoff matrices") explicitly encode evolutionary approximations regarding the rates and probabilities of particular amino acid mutations.

Another common series of scoring matrices, known as BLOSUM (Blocks Substitution Matrix), encodes empirically derived substitution probabilities. Variants of both types of matrices are used to detect sequences with differing levels of divergence, thus allowing users of BLAST or FASTA to restrict searches to more closely related matches or expand to detect more divergent sequences. Gap penalties account for the introduction of a gap - on the evolutionary model, an insertion or deletion mutation - in both nucleotide and protein sequences, and therefore the penalty values should be proportional to the expected rate of such mutations. The quality of the alignments produced therefore depends on the quality of the scoring function. It can be very useful and instructive to try the same alignment several times with different choices for scoring matrix and/or gap penalty values and compare the results. Regions where the solution is weak or non-unique can often be identified by observing which regions of the alignment are robust to variations in alignment parameters.

Other Biological Uses

Sequenced RNA, such as expressed sequence tags and full-length mRNAs, can be aligned to a sequenced genome to find where there are genes and get information about alternative splicing and RNA editing.

Sequence alignment is also a part of genome assembly, where sequences are aligned to find overlap so that *contigs* (long stretches of sequence) can be formed. Another use is SNP analysis, where sequences from different individuals are aligned to find single basepairs that are often different in a population.

Non-Biological Uses

The methods used for biological sequence alignment have also found applications in other fields, most notably in natural language processing and in social sciences, where the Needleman-Wunsch algorithm is usually referred to as Optimal matching.

Techniques that generate the set of elements from which words will be selected in natural-language generation algorithms have borrowed multiple sequence alignment techniques from bioinformatics to produce linguistic versions of computer-generated mathematical proofs. In the field of historical and comparative linguistics, sequence alignment has been used to partially automate the comparative method by which linguists traditionally reconstruct languages. Business and marketing research has also applied multiple sequence alignment techniques in analysing series of purchases over time.

Sequence Motifs

Frequently the primary structure encodes motifs that are of functional importance. Some examples of sequence motifs are: the C/D and H/ACA boxes of snoRNAs, Sm binding site found in spliceosomal RNAs such as U1, U2, U4, U5, U6, U12 and U3, the Shine-Dalgarno sequence, the Kozak consensus sequence and the RNA polymerase III terminator.

In genetics, a sequence motif is a nucleotide or amino-acid sequence pattern that is widespread and has, or is conjectured to have, a biological significance. For proteins, a sequence motif is distinguished from a structural motif, a motif formed by the three dimensional arrangement of amino acids, which may not be adjacent.

An Example is the N-glycosylation Site Motif: Asn, followed by anything but Pro, followed by either Ser or Thr, followed by anything but Prowhere the three-letter abbreviations are the conventional designations for amino acids. When a sequence motif appears in the exon of a gene, it may encode the "structural motif" of a protein; that is a stereotypical element of the overall structure of the protein. Nevertheless, motifs need not be associated with a distinctive secondary structure. "Noncoding" sequences are not translated into proteins, and nucleic acids with such motifs need not deviate from the typical shape (e.g. the "B-form" DNA double helix).

Outside of gene exons, there exist regulatory sequence motifs and motifs within the "junk", such as satellite DNA. Some of these are believed to affect the shape of nucleic acids, but this is only sometimes the case. For example, many DNA binding proteins that have affinity for specific DNA binding sites bind DNA in only its double-helical form. They are able to recognise motifs through contact with the double helix's major or minor groove.

Short coding motifs, which appear to lack secondary structure, include those that label proteins for delivery to particular parts of a cell, or mark them for phosphorylation. Within a sequence or database of sequences, researchers search and find motifs using computer-based techniques of sequence analysis, such as BLAST. Such techniques belong to the discipline of bioinformatics.

Motif Bioinformatics

Consider the *N*-glycosylation site motif mentioned above: *Asn, followed by anything but Pro, followed by either Ser or Thr, followed by anything but Pro.*

This pattern may be written as N{P}[ST]{P} where N = Asn, P = Pro, S = Ser, T = Thr; {X} means any amino acid except X; and [XY] means either X or Y.

The notation [XY] does not give any indication of the probability of X or Y occurring in the pattern. Observed probabilities can be graphically represented using sequence_logos. Sometimes patterns are defined in terms of a probabilistic model such as a hidden Markov model.

Motifs and Consensus Sequences

The notation [XYZ] means X or Y or Z, but does not indicate the likelihood of any particular match. For this reason, two or more patterns are often associated with a single motif: the defining pattern, and various typical patterns.

For example, the defining sequence for the IQ motif may be taken to be:

[FILV]Qxxx[RK]Gxxx[RK]xx[FILVWY]

where x signifies any amino acid, and the square brackets indicate an alternative.

Usually, however, the first letter is I, and both [RK] choices resolve to R. Since the last choice is so wide, the pattern IQxxxRGxxxR is sometimes equated with the IQ motif itself, but a more accurate description would be a *consensus sequence for the IQ motif.*

De Novo Computational Discovery of Motifs

There are software programs which, given multiple input sequences, attempt to identify one or more candidate motifs. One example is MEME, which generates statistical information for each candidate. Other algorithms include AlignAce, Amadeus, CisModule, FIRE, Gibbs Motif Sampler, PhyloGibbs, and Weeder. SCOPE is an ensemble motif finder that uses several algorithms simultaneously. PMS or PMS is another motif discovery program that is based on combinatorial approach.

There currently exist more than 100 publications with similar algorithms without a comprehensive benchmark so selecting one is not straightforward.

Discovery Through Evolutionary Conservation

Motifs have been discovered by studying similar genes in different species. For example, by aligning the amino acid sequences specified by the GCM (*glial cells missing*) gene in man, mouse and *D. melanogaster*, Akiyama and others discovered a pattern which they called the GCM motif. It spans about 150 amino acid residues, and begins as follows:

Here each . signifies a single amino acid or a gap, and each * indicates one member of a closely related family of amino acids.

The authors were able to show that the motif has DNA binding activity. PhyloGibbs and the Gibbs Motif Sampler are motif discovery algorithms that consider phylogenetic conservation.

Pattern Description Notations

Several notations for describing motifs are in use but most of them are variants of standard notations for regular expressions and use these conventions:

- there is an alphabet of single characters, each denoting a specific amino acid or a set of amino acids;
- a string of characters drawn from the alphabet denotes a sequence of the corresponding amino acids;
- any string of characters drawn from the alphabet enclosed in square brackets matches any one of the corresponding amino acids; e.g. [abc] matches any of the amino acids represented by a or b or c.

The fundamental idea behind all these notations is the matching principle, which assigns a meaning to a sequence of elements of the pattern notation:

a sequence of elements of the pattern notation matches a sequence of amino acids if and only if the latter sequence can be partitioned into subsequences in such a way that each pattern element matches the corresponding subsequence in turn.

Thus the pattern [AB] [CDE] F matches the six amino acid sequences corresponding to ACF, ADF, AEF, BCF, BDF, and BEF. Different pattern description notations have other ways of forming pattern elements. One of these notations is the PROSITE notation, described in the following subsection.

PROSITE Pattern Notation

The PROSITE notation uses the IUPAC one-letter codes and conforms to the above description with the exception that a concatenation symbol, '-', is used between pattern elements, but it is often dropped between letters of the pattern alphabet. PROSITE allows the following pattern elements in addition to those described previously:

- The lower case letter 'x' can be used as a pattern element to denote any amino acid.
- A string of characters drawn from the alphabet and enclosed in braces (curly brackets) denotes any amino acid except for those in the string. For example, {ST} denotes any amino acid other than S or T.
- If a pattern is restricted to the N-terminal of a sequence, the pattern is prefixed with '<'.
- If a pattern is restricted to the C-terminal of a sequence, the pattern is suffixed with '>'.
- The character '>' can also occur inside a terminating square bracket pattern, so that S[T>] matches both "ST" and "S>".
- If e is a pattern element, and m and n are two decimal integers with m <= n, then:
    - o e(m) is equivalent to the repetition of e exactly m times;
    - o e(m,n) is equivalent to the repetition of e exactly k times for any integer k satisfying: m <= k <= n.

Some examples:

- x(3) is equivalent to x-x-x.
- x(2,4) matches any sequence that matches x-x or x-x-x or x-x-x-x.

The signature of the C2H2-type *zinc finger* domain is:

- C-x(2,4)-C-x(3)-[LIVMFYWC]-x(8)-H-x(3,5)-H

Relation to Secondary and Tertiary Structure

The primary structure of a biological polymer to a large extent determines the three-dimensional shape known as the tertiary structure, but nucleic acid and protein folding are so complex that knowing the primary structure often doesn't help either to deduce the shape or to predict localized secondary structure, such as the formation of loops or helices.

However, knowing the structure of a similar homologous sequence (for example a member of the same protein family) can unambiguously identify the tertiary structure of the given sequence. Sequence families are often determined by sequence clustering, and structural genomics projects aim to produce a set of representative structures to cover the *sequence space* of possible non-redundant sequences.

Biomolecular structure is the structure of biomolecules, mainly proteins and the nucleic acids DNA and RNA. The structure of these molecules is frequently decomposed into primary structure, secondary structure, tertiary structure, and quaternary structure. The scaffold for this structure is provided by secondary structural elements which are hydrogen bonds within the molecule. This leads to several recognisable "domains" of protein structure and nucleic acid structure, including secondary structure like hairpin loops, bulges and internal loops for nucleic acids, and alpha helices and beta sheets for proteins.

The terms primary, secondary, tertiary, and quaternary structure were first coined by Kaj Ulrik Linderstrøm-Lang in his 1951 Lane Medical Lectures at Stanford University.

In biochemistry, the Primary structure of a biological molecule is the exact specification of its atomic composition and the chemical bonds connecting those atoms (including stereochemistry). For a typical unbranched, un-crosslinked biopolymer (such as a molecule of DNA, RNA or typical intracellular protein), the primary structure is equivalent to specifying the sequence of its monomeric subunits, e.g., the nucleotide or peptide sequence.

Primary structure is sometimes mistakenly termed *primary sequence*, but there is no such term, as well as no parallel concept of secondary or tertiary sequence. By convention, the primary structure of a protein is reported starting from the amino-terminal (N) end to the carboxyl-terminal (C) end, while the primary structure of DNA or RNA molecule is reported from the 5' end to the 3' end.

The primary structure of a nucleic acid molecule refers to the exact sequence of nucleotides that comprise the whole molecule. Frequently the primary structure encodes motifs that are of functional importance. Some examples of sequence motifs are: the C/D and H/ACA boxes of snoRNAs, Sm binding site found in spliceosomal RNAs such as U1, U2, U4, U5, U6, U12 and U3, the Shine-Dalgarno sequence, the Kozak consensus sequence and the RNA polymerase III terminator.

Secondary Structure

In biochemistry and structural biology, secondary structure is the general three-dimensional form of *local segments* of biopolymers such as proteins and nucleic acids (DNA/RNA). It does not, however, describe specific atomic positions in three-dimensional space, which are considered to be tertiary structure.

Secondary structure is formally defined by the hydrogen bonds of the biopolymer, as observed in an atomic-resolution structure. In proteins, the secondary structure is defined by patterns of hydrogen bonds between backbone amide and carboxyl groups (sidechain-mainchain and sidechain-sidechain hydrogen bonds are irrelevant), where the DSSP definition of a hydrogen bond is used. In nucleic acids, the secondary structure is defined by the hydrogen bonding between the nitrogenous bases.

For proteins, however, the hydrogen bonding is correlated with other structural features, which has given rise to less formal definitions of secondary structure. For example, residues in protein helices generally adopt backbone dihedral angles in a particular region of the Ramachandran plot; thus, a segment of residues with such dihedral angles is often called a "helix", regardless of whether it has the correct hydrogen bonds.

Many other less formal definitions have been proposed, often applying concepts from the differential geometry of curves, such as curvature and torsion. Least formally, structural biologists solving a new atomic-resolution structure will sometimes assign its secondary structure "by eye" and record their assignments in the corresponding PDB file. The secondary structure of a nucleic acid molecule refers to the basepairing interactions within a single molecule or set of interacting molecules. The secondary structure of biological RNA's can often be uniquely decomposed into stems and loops. Frequently these elements, or combinations of them, can be further classified, for example, tetraloops, pseudoknots and stem-loops. There are many secondary structure elements of functional importance to biological RNA's; some famous examples are the

Rho-independent terminator stem-loops and the tRNA cloverleaf. There is a minor industry of researchers attempting to determine the secondary structure of RNA molecules. Approaches include both experimental and computational methods. In biochemistry and structural biology, secondary structure is the general three-dimensional form of *local segments* of biopolymers such as proteins and nucleic acids (DNA/RNA). It does not, however, describe specific atomic positions in three-dimensional space, which are considered to be tertiary structure.

Secondary structure can be formally defined by the hydrogen bonds of the biopolymer, as observed in an atomic-resolution structure. In proteins, the secondary structure is defined by the patterns of hydrogen bonds between backbone amide and carboxyl groups.

In nucleic acids, the secondary structure is defined by the hydrogen bonding between the nitrogenous bases. The hydrogen bonding patterns may be significantly distorted, which makes an automatic determination of secondary structure difficult.

The secondary structure may be also defined based on the regular pattern of backbone dihedral angles in a particular region of the Ramachandran plot; thus, a segment of residues with such dihedral angles may be called a helix, regardless of whether it has the correct hydrogen bonds. The secondary structure may be also provided by crystallographers in the corresponding PDB file.

The rough secondary-structure content of a biopolymer (e.g., "this protein is 40% α-helix and 20% β-sheet.") can often be estimated spectroscopically. For proteins, a common method is far-ultraviolet (far-UV, 170-250 nm) circular dichroism.

A pronounced double minimum at 208 and 222 nm indicate α-helical structure, whereas a single minimum at 204 nm or 217 nm reflects random-coil or β-sheet structure, respectively. A less common method is infrared spectroscopy, which detects differences in the bond oscillations of amide groups due to hydrogen-bonding. Finally, secondary-structure contents may be estimated accurately using the chemical shifts of an unassigned NMR spectrum.

Secondary structure was introduced by Kaj Ulrik Linderstrøm-Lang at Stanford in 1952.

Secondary structure in proteins consists of local inter-residue interactions mediated by hydrogen bonds, or not. The most common secondary structures are alpha helices and beta sheets. Other helices, such as the 3_{10} helix and π helix, are calculated to have energetically favourable hydrogen-bonding patterns but are rarely if ever observed in natural proteins except at the ends of α helices due to unfavourable backbone packing in the centre of the helix. Other extended structures such as the polyproline helix and alpha sheet are rare in native state proteins but are often hypothesized as important protein folding intermediates. Tight turns and loose, flexible loops link the more "regular" secondary structure elements. The random coil is not a true secondary structure, but is the class of conformations that indicate an absence of regular secondary structure.

Amino acids vary in their ability to form the various secondary structure elements. Proline and glycine are sometimes known as "helix breakers" because they disrupt the regularity of the α helical backbone conformation; however, both have unusual conformational abilities and are commonly found in turns.

Amino acids that prefer to adopt helical conformations in proteins include methionine, alanine, leucine, glutamate and lysine ("MALEK" in amino-acid 1-letter codes); by contrast, the large aromatic residues (tryptophan, tyrosine and phenylalanine) and -branched amino acids (isoleucine, valine, and threonine) prefer to adopt β-strand conformations.

However, these preferences are not strong enough to produce a reliable method of predicting secondary structure from sequence alone.

The DSSP Code

The Dictionary of Protein Secondary Structure, in short DSSP, is commonly used to describe the protein secondary structure with single letter codes. The secondary structure is assigned based on hydrogen bonding patterns as those initially proposed by Pauling et al. in 1951 (before any protein structure had ever been experimentally determined). There are eight types of secondary structure that DSSP defines:

- G = 3-turn helix (3_{10} helix). Min length 3 residues.
- H = 4-turn helix (α helix). Min length 4 residues.
- I = 5-turn helix (π helix). Min length 5 residues.
- T = hydrogen bonded turn (3, 4 or 5 turn)
- E = extended strand in parallel and/or anti-parallel β-sheet conformation. Min length 2 residues.
- B = residue in isolated β-bridge (single pair β-sheet hydrogen bond formation)
- S = bend (the only non-hydrogen-bond based assignment).

Amino acid residues which are not in any of the above conformations are assigned as the eighth type 'Coil': often codified as ' ' (space), C (coil) or '-' (dash).

The helices (G,H and I) and sheet conformations are all required to have a reasonable length. This means that 2 adjacent residues in the primary structure must form the same hydrogen bonding pattern. If the helix or sheet hydrogen bonding pattern is too short they are designated as T or B, respectively. Other protein secondary structure assignment categories exist (sharp turns, Omega loops etc.), but they are less frequently used.

Protein Secondary-Structure Prediction

Predicting protein tertiary structure from only its amino acid sequence is a very challenging problem, but using the simpler secondary structure definitions is more tractable and has been the focus for research for a long time.

Although, the 8-state DSSP code is already a simplification from the continuous variation of hydrogen bonding patterns present in a protein the majority of secondary prediction methods simplify further to the three dominant states: Helix, Sheet and Coil. How the conversion is made from 8- to 3-state varies between methods.

Early methods of secondary-structure prediction were based on the helix- or sheet-forming propensities of individual amino acids, sometimes coupled with rules for estimating the free energy of forming secondary structure elements. Such methods were typically ~60% accurate in predicting which of the three states (helix/sheet/coil) a residue adopts.

A significant increase in accuracy (to nearly ~80%) was made by exploiting multiple sequence alignment; knowing the full distribution of amino acids that occur at a position (and in its vicinity, typically ~7 residues on either side) throughout evolution provides a much better picture of the structural tendencies near that position. For illustration, a given protein might have a glycine at a given position, which by itself might suggest a random coil there. However, multiple sequence alignment might reveal that helix-favouring amino acids occur at that position (and nearby positions) in 95% of homologous proteins spanning nearly a billion years of evolution. Moreover, by examining the average hydrophobicity at that and nearby positions, the same alignment might also suggest a pattern of residue solvent accessibility consistent with an α-helix. Taken together, these factors would suggest that the glycine of the original protein adopts α-helical structure, rather than random coil.

Several types of methods are used to combine all the available data to form a 3-state prediction, including neural networks, hidden Markov models and support vector machines. Modern prediction methods also provide a confidence score for their predictions at every position. Secondary-structure prediction methods are continuously benchmarked, e.g., in the EVA experiment. Based on ~270 weeks of testing, the most accurate methods at present are PSIPRED, SAM, PORTER, PROF and SABLE. Interestingly, it does not seem to be possible to improve upon these methods by taking a consensus of them. The chief area for improvement appears to be the prediction of β-strands; residues confidently predicted as β-strand are likely to be so, but the methods are apt to overlook some β-strand segments (false negatives). There is likely an upper limit of ~90% prediction accuracy overall, due to the idiosyncrasies of the standard method (DSSP) for assigning secondary-structure classes (helix/strand/coil) to PDB structures, against which the predictions are benchmarked.

Accurate secondary-structure prediction is a key element in the prediction of tertiary structure, in all but the simplest (homology modelling) cases. For example, a confidently predicted pattern of six secondary structure elements βαββαβ is the signature of a ferredoxin fold.

β Sheet

β Sheets are formed by H bonds between an average of 5–10 consecutive amino acids in one portion of the chain with another 5–10 farther down the chain. The interacting regions may be adjacent, with a short loop in between, or far apart, with other structures in between.

Every chain may run in the same direction to form a parallel sheet, every other chain may run in the reverse chemical direction to form an anti parallel sheet, or the chains may be parallel and anti parallel to form a mixed sheet. The pattern of H bonding is different in the parallel and anti parallel configurations.

Each amino acid in the interior strands of the sheet forms two H bonds with neighbouring amino acids, whereas each amino acid on the outside strands forms only one bond with an interior strand. Looking across the sheet at right angles to the strands, more distant strands are rotated slightly counterclockwise to form a left-handed twist. The Cα atoms alternate above and below the sheet in a pleated structure, and the R side groups of the amino acids alternate above and below the pleats. The Φ and Ø angles of the amino acids in sheets vary considerably in one region of the Ramachandran plot. It is more difficult to predict the location of β sheets than of α helices. The situation improves some what when the amino acid variation in multiple sequence alignments is taken into account.

Loop

Loops are regions of a protein chain that are (1) between α helices and β sheets, (2) of various lengths and three-dimensional configurations, and (3) on the surface of the structure. Hairpin loops that represent a complete turn in the polypeptide chain joining two antiparallel β strands may be as short as two amino acids in length. Loops interact with the surrounding aqueous environment and other proteins. Because amino acids in loops are not constrained by space and environment as are amino acids in the core region, and do not have an effect on the arrangement of secondary structures in the core, more substitutions, insertions, and deletions may occur.

Thus, in a sequence alignment, the presence of these features may be an indication of a loop. The positions of introns in genomic DNA sometimes correspond to the locations of loops in the encoded protein. Loops also tend to have charged and polar amino acids and are frequently a component of active sites. A detailed examination of loop structures has shown that they fall into distinct families.

Coils

A region of secondary structure that is not a helix, a β sheet, or a recognisable turn is commonly referred to as a coil.

Chapter 5

Chemical Synthesis

In chemistry, chemical synthesis is purposeful execution of chemical reactions to get a product, or several products. This happens by physical and chemical manipulations usually involving one or more reactions. In modern laboratory usage, this tends to imply that the process is reproducible, reliable, and established to work in multiple laboratories.

A chemical synthesis begins by selection of compounds that are known as reagents or reactants. Various reaction types can be applied to these to synthesize the product, or an intermediate product. This requires mixing the compounds in a reaction vessel such as a chemical reactor or a simple round-bottom flask. Many reactions require some form of work-up procedure before the final product is isolated.

The amount of product in a chemical synthesis is the reaction yield. Typically, chemical yields are expressed as a weight in grams or as a percentage of the total theoretical quantity of product that could be produced. A side reaction is an unwanted chemical reaction taking place that diminishes the yield of the desired product. The word *synthesis* in the present day meaning was first used by the chemist Adolph Wilhelm Hermann Kolbe.

Strategies

Many strategies exist in chemical synthesis that go beyond converting reactant A to reaction product B. In cascade reactions multiple chemical trans-formations take place within a single reactant, in multi-component reactions up to 11 different reactants form a single reaction product and in a telescopic synthesis one reactant goes through multiple transformations without isolation of intermediates.

Organic Synthesis

Organic synthesis is a special branch of chemical synthesis dealing with the synthesis of organic compounds. In the total synthesis of a complex product it may take multiple steps to synthesize the product of interest, and inordinate amounts of time.

Skill in organic synthesis is prized among chemists and the synthesis of exceptionally valuable or difficult compounds has won chemists such as Robert Burns Woodward the Nobel Prize for Chemistry. If a chemical synthesis starts from basic laboratory compounds and yields something new, it is a purely synthetic process. If it starts from a product isolated from plants or animals and then proceeds to new compounds, the synthesis is described as a semisynthetic process.

Organic synthesis is a special branch of chemical synthesis and is concerned with the construction of organic compounds via organic reactions. Organic molecules can often contain a higher level of complexity compared to purely inorganic compounds, so the synthesis of organic compounds has developed into one of the most important branches of organic chemistry. There are two main areas of research fields within the general area of organic synthesis: *total synthesis* and *methodology*.

Total Synthesis

A total synthesis is the complete chemical synthesis of complex organic molecules from simple, commercially available (petrochemical) or natural precursors. In a *linear* synthesis—often adequate for simple structures—several steps are performed one after another until the molecule is complete.

The chemical compounds made in each step are usually deemed *synthetic intermediates.* For more complex molecules, a different approach may be preferable: convergent synthesis involves the individual preparation of several "pieces" (key intermediates), which are then combined to form the desired product. Robert Burns Woodward, who received the 1965 Nobel Prize for Chemistry for several total syntheses (e.g., his 1954 synthesis of strychnine), is regarded as the father of modern organic synthesis. Some latter-day examples include Wender's, Holton's, Nicolaou's and Danishefsky's synthesis of Taxol.

In organic chemistry, a total synthesis is, in principle, the complete chemical synthesis of complex organic molecules from simpler pieces, usually without the aid of biological processes.

In practice, these simpler pieces are commercially available in bulk and semi-bulk quantities, and are often petrochemical precursors. Sometimes bulk natural products (e.g. sugars) are used as starting materials and it is assumed that these have or can be synthesised from their constituent elements.

The target molecules can be natural products (biomolecules), medicinally important active ingredients, or organic compounds of theoretical interest in chemistry or biology. A new route for synthesis is developed in the course of the investigation, and the route may be the first one to be developed for the substance.

The first demonstration of organic total synthesis was Friedrich Wöhler's synthesis of urea in 1828, which demonstrated that organic molecules can be produced from inorganic precursors, and the first commercialised total synthesis was Gustaf Komppa's synthesis and industrial production of camphor in 1903.

Early efforts focused on building chemicals which were extracted from biological sources, and using them to verify their biological activities—in this fashion, total synthesis was associated with disproving the existence of a vital force. As the sophistication of biology and chemistry became more understood, the primary purpose for total synthesis changed, although some total synthesis continues to be a tool for biological verification, for example if a compound contains ambiguous stereochemistry or to create analogs ("chemical mutations") to directly assess or improve the mechanism of biological activity.

Today, total synthesis is often justified as a playground for the development of new chemical reactions and routes, and highlights the sophistication of modern synthetic organic chemistry. Sometimes total synthesis inspires the development of novel mechanisms, catalysts, or techniques. Finally since a total synthesis project will often span a variety of reactions, it serves to prepare chemists for pursuits in process chemistry, where encyclopedic knowledge of chemical reactions and a strong and accurate sense of chemical intuition are necessary.

Formal Synthesis

A formal synthesis describes not the synthesis of the desired end product but the synthesis of a compound that is already known from the literature to be a precursor to that desired end product. If it is known from the literature that B can be converted to C then a novel route from compound A to compound B is a formal proof that A can also give access to C.

Examples

Classical examples are the total syntheses of cholesterol, cortisone, strychnine, lysergic acid, reserpine, chlorophyll, colchicine, vitamin B_{12} and prostaglandin F-2a by the Nobel Prize winner Robert Burns Woodward between 1945 and 1976.

Another classic is quinine total synthesis that has a history spanning 150 years. In some cases the structures of the molecules assigned by spectroscopic methods have been shown to be wrong, when synthesised.

Elias James Corey won the Nobel Prize in Chemistry in 1990 for lifetime achievement in total synthesis and the development of retrosynthetic analysis. In 2005 the group of Corey published an aflatoxin total synthesis and in 2006 an oseltamivir total synthesis.

Methodology and Applications

Each step of a synthesis involves a chemical reaction, and reagents and conditions for each of these reactions need to be designed to give a good yield and a pure product, with as little work as possible. A method may already exist in the literature for making one of the early synthetic intermediates, and this method will usually be used rather than "trying to reinvent the wheel".

However most intermediates are compounds that have never been made before, and these will normally be made using general methods developed by methodology researchers.

To be useful, these methods need to give high yields, and to be reliable for a broad range of substrates. For practical applications, additional hurdles include industrial standards of safety and purity. Methodology research usually involves three main stages: *discovery*, *optimisation*, and studies of *scope and limitations*.

The *discovery* requires extensive knowledge of and experience with chemical reactivities of appropriate reagents. *Optimisation* is where one or two starting compounds are tested in the reaction under a wide variety of conditions of temperature, solvent, reaction time, etc., until the optimum conditions for product yield and purity are found. Finally, the researcher tries to extend the method to a broad range of different starting materials, to find the scope and limitations. Total synthesis are sometimes used to showcase the new methodology and demonstrate its value in a real-world application. Such applications involved major industries focused especially on polymers (and plastics) and on pharmaceuticals.

Stereoselective Synthesis

Most complex natural products are chiral, and the bioactivity of chiral molecules varies with the enantiomer. Traditional total syntheses targeted racemic mixtures, i.e., as an equal mixture of both possible enantiomers. The racemic mixture might then be separated via chiral resolution.

In the latter half of the twentieth century, chemists began to develop methods of stereoselective catalysis and kinetic resolution whereby reactions could be directed to produce only one enantiomer rather than a racemic mixture. Early examples include Sharpless epoxidation (K. Barry Sharpless) and stereoselective hydrogenation (William S. Knowles and Ryôji Noyori). For their achievement, these workers went on to share the Nobel Prize in Chemistry in 2001. Such reactions gave chemists a much wider choice of enantiomerically pure molecules to start from, where previously only natural starting materials could be used. Using techniques pioneered by Robert B. Woodward and new developments in synthetic methodology, chemists became more able to take simple molecules through to more complex molecules without unwanted racemisation, by understanding stereocontrol.

This allowed the final target molecule to be synthesised as one pure enantiomer without any resolution being necessary. Such techniques are referred to as stereoselective synthesis.

Enantioselective synthesis, also called chiral synthesis, asymmetric synthesis or stereoselective synthesis, is organic synthesis that selectively produces one or more new and desired elements of chirality. This methodology is important in the field of pharmaceuticals because the different enantiomers or diastereomers of a molecule often have different biological activity.

Approaches

There are three main approaches to enantioselective synthesis:

- chiral pool synthesis
- chiral auxiliaries
- enantioselective catalysis

It is often advantageous to combine two or all of these approaches.

Chiral Pool Synthesis

Chiral pool synthesis is the easiest approach: A chiral starting material is manipulated through successive reactions using achiral reagents that retain its chirality to obtain the desired target molecule. This is especially attractive for target molecules having the similar chirality to a relatively inexpensive naturally occurring building-block such as a sugar or amino acid. However, the number of possible reactions the molecule can undergo is restricted, and tortuous synthetic routes may be required. Also, this approach requires a stoichiometric amount of the enantiopure starting material, which may be rather expensive if not occurring in nature, whereas chiral catalysis requires only a catalytic amount of chiral material.

Enantioselective Induction

What many strategies in chiral synthesis have in common is asymmetric induction. The aim is to make enantiomers into diastereomers, since diastereomers have different reactivity, but enantiomers do not.

To make enantiomers into diastereomers, the reagents or the catalyst need to be incorporated with an enantiopure chiral centre. The reaction will now proceed differently for different enantiomers, because the transition state of the reaction can exist in two diastereomers with respect to the enantiopure centre, and these diastereomers react differently. Enantioselective induction can also occur intramolecularly when given a chiral starting material. This enantioinduction can be exploited, especially when the goal is to make several consecutive chiral centres to give a specific enantiomer of a specific diastereomer. An aldol reaction, for example, is inherently diastereoselective; if the aldehyde is enantiopure, the resulting aldol adduct is diastereomerically and enantiomerically pure.

Chiral Auxiliary

One enantioinduction strategy is the use of a chiral auxiliary, which forms an adduct to the starting materials and physically blocks the other trajectory for attack, leaving only the desired trajectory open.

Assuming the chiral auxiliary is enantiopure, the different trajectories are not equivalent, but diastereomeric. The auxiliary shares problems similar to protecting groups; like protecting groups, auxiliaries require a reaction step to add and another to remove, increasing cost and decreasing yield.

Biocatalysis & Organocatalysis

Biocatalysis makes use of enzymes to effect chemical reagents stereoselectively. Some small organic molecules can also be used to help accelerate the desired reaction; this method is known as organocatalysis. If the organic molecule is chiral, it may react preferentially with a particular enantiomer or diastereomer of substrate.

Alternatives

Apart from enantioselective synthesis, racemic mixtures of compounds can be resolved by various techniques in chiral resolution. Where the cost in time and money of making such racemic mixtures is low, or if both enantiomers may find use, this approach may remain cost-effective.

Planning the Synthesis

Chirality must be introduced to the substance first. Then, it must be maintained. Care needs to be taken when planning the synthesis: a chemical change that makes the substance isotropic renders it achiral. This process is called epimerization.

For example, a S_N1 substitution reaction converts a molecule that is chiral by merit of non-planarity into a planar molecule, which has no handedness. (To visualise, draw the outlines of both of your hands on paper, and cut the images out. You can now superimpose the images, even if the hands themselves do not superimpose.) In a S_N2 substitution reaction, on the other hand, the chirality inverts, i.e., when you start with a right-handed mixture, you'll end up with left-handed one. (A visualization could be inverting an umbrella. The mechanism looks just the same.)

Synthesis Design

Elias James Corey brought a more formal approach to synthesis design, based on retrosynthetic analysis, for which he won the Nobel Prize for Chemistry in 1990. In this approach, the research is planned backwards from the product, using standard rules. The steps are shown using retrosynthetic arrows (drawn as Ò!), which in effect means "is made from". Computer programs have been written for designing a synthesis based on sequences of generic "half-reactions".

Chiral Lewis Acid

Chiral Lewis acids (CLAs) are a novel class of Lewis acid catalyst used in enantioselective asymmetric synthesis reactions which produce optically active products from optically inactive or impure starting materials. This type of preferential formation of one enantiomer or diastereomer over the other is formally an asymmetric induction.

The electron accepting atom of CLAs are typically metal, such as indium, zinc, lithium, aluminium, titanium, or boron – and the chiral ligands employed for synthesizing CLAs most often have multiple Lewis basic sites (often diols or dinitrogen structures) to allow forming a ring structure involving the metal atom.

While the ability of achiral Lewis acids to promote the synthesis of racemic mixtures in myriad different reactions has been known for quite some time, starting in the 1960s, CLAs have been used to induce the enantioselectivity in these reactions. Common reaction types include Diels-Alder reactions, the ene reaction, [2+2] cycloaddition reactions, hydrocyanation of aldehydes, and most notably, Sharpless expoxidations. The enantioselectivity of CLAs derives from their ability to perturb the free energy barrier along the reaction coordinate pathway that leads to either the *R*- or *S*- enantiomer. Ground state diastereomers and enantiomers are of equal energy in the ground state, and when reacted with an achiral lewis acid, their diastereomeric intermediates, transition states, and products are also of equal energy.

This leads to the production of racemic mixtures of products. However, when a CLA is utilised in the same reaction, the energetic barrier of formation of one diastereomer is less than that of another – the reaction is under kinetic control. If the difference in the energy barriers between the diastereomeric transition states are of sufficient magnitude, and high enantiomeric excess of one isomer should be observed .

Review Details

A 2011 systematic review from The Cochrane Library analysed 48 studies conducted between 1965 and 2009 and included 65,508 participants. All studies reduced or modified participants' dietary fat or cholesterol for at least six months by at least 30 percent.

It was found that reducing saturated fat by reducing and/ or modifying dietary fat reduced the risk of having a cardiovascular event, such as heart attack, stroke and unplanned heart surgery, by 14 percent. Of the 65,508 participants, 7 percent had a cardiovascular event.

> *"The findings are suggestive of a small but potentially important reduction in cardiovascular risk on modification of dietary fat, but not reduction of total fat, in longer trials. Lifestyle advice to all those at risk of cardiovascular disease and to lower risk population groups, should continue to include permanent reduction of dietary saturated fat and partial replacement by unsaturates. The ideal type of unsaturated fat is unclear". In a summary it goes on to say "there are no clear health benefits of replacing saturated fats with starchy foods".*

A 2010 systematic review and meta-analysis of randomized controlled trials commissioned by the National Heart, Lung, and Blood Institute found that substitution of poly-unsaturated fat for saturated fat led to a 19% reduction in Coronary Heart Disease. In 2010, a meta-analysis of prospective cohort studies supported by the National Dairy Council including 348,000 subjects found no statistically significant relationship between cardiovascular disease and dietary saturated fat.

In a subordinate meta-analysis to a 2009 risk assessment supported by the Centres for Disease Control and Prevention (CDC), an ischemic heart disease mortality residual risk of 1.05 was determined for the 30-44 year age group for each 1% isocaloric reduction in PUFA in favour of SFA. Higher age groups had lower residual risks. In 2009, a systematic review supported by the Heart and Stroke Foundation of Canada of prospective cohort studies or randomized trials concluded that there was "insufficient evidence of association" between intake of saturated fatty acids and coronary heart disease, and pointed to strong evidence for protective factors such as vegetables and a Mediterranean diet and harmful factors such as trans fats and foods with a high glycemic index. A 2009 meta-analysis supported by the NHBLI focused on the effects of trans-fats also applied the same techniques to studies focused on SFA, MUFA and PUFA, in order to assess the relative risk of substituting trans-fats with other fats. Risks were assessed from two classes of data - risk factors in trials, and disease outcomes in cohort studies.

In order of decreasing risk, the fat types were TFA, SFA, MUFA and PUFA. From the trials analysis, it was found that for partially hydrogenated vegetable oil with 20% trans-fats at consumption rate of 7.5% of daily calories, replacement with butter would decrease risk by 2.7% and replacement with canola oil would reduce risk by 9.9%. From the cohort analysis, butter would have little net effect (0.5% increased risk) while soybean and canola oil would have the largest reduction in risk (19.0 and 21.8%, respectively). A 2009 meta-analysis of cohort studies and randomized controlled trials from the University of Otago, New Zealand, found that the results of the cohort studies showed no association between coronary heart disease mortality nor total coronary heart disease events when saturated fat was increased in place of carbohydrates. In the randomized controlled trials where the ratio of polyunsaturated fat to saturated fat had been increased, the risk of coronary heart disease fatality was not changed, but the risk of total coronary heart disease events was decreased. The analysis was critical of the limitations of the primary evidence available, considering the evidence from cohort studies to be mostly unreliable.

A 2009 pooled analysis of 11 cohort studies supported by the National Heart, Lung, and Blood Institute, National Institutes of Health, the Danish Heart Foundation and the Danish Medical Research Council found that substituting polyunsaturated fatty acids in place of saturated fatty acids at a rate of 5% of energy intake led to 13% decrease in coronary events and a 26% decrease in coronary deaths.

The American Dietetic Association's 2008 systematic review found that a diet with 25%-35% total fat but less than 7% saturated fat and trans fat lowers the risk of coronary heart disease. A 2003 meta-analysis of prevention dietetic studies found the benefit of a modified fat diet in primary prevention is potentially significant if it is maintained for a sufficient length of time.

A 1999 review found that substitution of the fat from one ounce of nuts for equivalent energy from saturated fat was associated with a 45% reduction in risk of coronary heart disease.

A 1994 meta-analysis from the University of Sydney of the effect on coronary events and total mortality in dietary intervention trials found a 6% reduction in mortality and 13% reduction in events.

The dietary advice differed between the trials reviewed, but most prescribed a reduced saturated fat and cholesterol intake with partial replacement by polyunsaturated oils. The review noted that a large controlled trial "may no longer be ethical."

Some meta-analyses have found a significant relationship between saturated fat and serum cholesterol levels, and serum cholesterol levels and cardiovascular disease. However, the ratio of total to high density lipoprotein cholesterol is considered a better indicator and some saturated fats (lauric acid and stearic acid) improve the ratio.

Mainstream Authorities

Specialist/Professional Textbooks

The 2009 *European Society of Cardiology Textbook of Cardiovascular Medicine* states that in cohort studies the positive relationship between fat intake and CVDs was linked to their saturated fatty acid content. 2007's *Cardiovascular Prevention and Rehabilitation* states that large epidemiological studies have shown consistent associations between the intake of saturated fatty acids and CHD mortality.

According to the 2007 *Critical Pathways in Cardiovascular Medicine*, substituting unsaturated fat for saturated fat may lower LDL cholesterol without simultaneously lowering HDL cholesterol. This dietary principle partly underlies the Mediterranean style of diet, which has been associated with reduced cardiovascular event rates in two randomized controlled trials. The 2003 second edition of *Evidence-based Cardiology* in 'PartII: Prevention of cardiovascular diseases' recommends a low intake of SFA, less than 7% of daily calories, and intake of foods rich in myristic and palmitic acids should be especially reduced. The recommendation was evaluated to be supported by the best grade of available evidence.

Lipid Hypothesis of Atherogenesis

Early Studies and Origin

In 1913, a study by Nikolai Anitschkow showed that feeding rabbits cholesterol could induce symptoms similar to atherosclerosis, suggesting a role for cholesterol in atherogenesis.

In 1951, Duff and McMillian formulated the lipid hypothesis in its modern form in a review which appeared in the *American Journal of Medicine.* One of the best known early modern proponents that saturated fats and cholesterol in the blood are the cause of heart disease was Ancel Keys, whose first paper on the topic was published in 1953, and whose book *Eat Well and Stay Well*, helped the issue gain popular awareness.

Cholesterol Controversy of Atherogenesis

During the 1970s and 1980s, some researchers and practitioners considered the lipid hypothesis as unverified due to the lack of proof at that time that lowering blood cholesterol levels resulted in decreased risk for atherosclerosis. Some skeptics were questioning its validity by arguing that the studies supporting it were flawed.

This discussion is also referred to as the "cholesterol controversy". Predictions were made at the time that further research during the 1980s and 1990s would help settle this controversy. However, even after the Coronary Primary Prevention Trial and the NIH Consensus Conference in the mid 1980s, criticism persisted in a minority of the scientific community questioning the statistical and mechanistic significance of the associations measured in the trials and the conclusions of the panel. In the following years, studies with lipid- and cholesterol-lowering drugs such as statins provided further associative evidence in support of the lipid hypothesis. However, a meta-analysis of cholesterol-lowering trials found that trials that were supportive of the lipid hypothesis were cited almost six times as often as those that were not, and although there was a similar number of trials unsupportive of the hypothesis, none of them were cited after 1970, some of the supportive reviews also exclude and ignore certain trials which were less favourable to the hypothesis; this meta-analysis, considering the less-cited trials, found that mortality was not decreased by lowering cholesterol, and that the lowering of cholesterol was unlikely to prevent coronary heart disease.

One proposed mechanism by which statins reduce coronary heart disease is by working as vitamin D analogues. Further studies were designed in the hope of settling the controversy. The continuing consensus regards the lipid hypothesis as proven; but disagreement still persists among groups who argue that it is based on associations and misrepresented or over-interpreted data, and has not been shown as a scientifically validated causal mechanism.

A documentary film titled *Fat Head* (2009) covers, in the second half of the movie, the origin and the propagation of the lipid hypothesis. The film chronicles the claims made by the main proponents of the lipid hypothesis, how their work influenced the nutrition guidelines published by the U.S. government, and the alleged lack of scientifically valid statistical evidence for those claims.

Other Lipid Hypotheses

Alzheimer's Lipid Hypothesis and Cholesterol Controversy

A new "cholesterol controversy" is emerging out of the research of the pathogenesis of Alzheimer's disease (AD). A "lipid hypothesis of AD" has been proposed that suggests a connection between cholesterol and changes in brain lipids found in Alzheimer-type neuropathology.

Lipid Hypothesis of Osteoporosis

The "lipid hypothesis of osteoporosis" postulates that lipids involved in causing heart disease also contribute to causing osteoporosis. Osteoporosis is characterised by a decrease of bone marrow cells, or osteoblasts, and an increase of fat cells, or adipocytes. The formation of osteoblasts from preosteoblasts is reduced by oxidized lipids and in mice fed with a high fat diet. Observations from this model suggest that LDL oxidation products can cause osteoporosis through changing the developmental fate of bone cells leading to a reduced number of osteoblasts and increased numbers of fat cells.

Lipid Hypothesis of Cold Tolerance

In plants and microbes, changes in the lipid composition of cell membranes have been linked to cold tolerance. The enhanced resistance to cold treatment appears to be caused by an increased amount of fatty acid desaturases produced under cold stress transforming saturated into unsaturated fatty acids in the membrane. This effect can be reproduced artificially in genetically engineered plants. The changes in membrane lipid composition lead to a higher membrane fluidity, thus keeping the membrane from "freezing" at low temperatures. This "lipid hypothesis of cold tolerance" is less well supported in animals. In fruit flies, cold acclimation does not coincide with a reduced amount of saturated fatty acids, and recent genetic studies on a nematode indicate that the mechanisms involved in cold adaptation in animals may be different from those in plants and microbes. A meta-analysis published in 2003 found a significant positive relationship in both control and cohort studies between saturated fat and breast cancer. However two subsequent reviews have found weak or insignificant associations of saturated fat intake and breast cancer risk, and note the prevalence of confounding factors. A systematic literature review published by the World Cancer Research Fund and the American Institute for Cancer Research in 2007 found limited but consistent evidence for a positive relationship between animal fat and colorectal cancer.

A meta-analysis published in 2001 found a statistically significant positive relationship between saturated fat and ovarian cancer.

Some researchers have indicated that serum myristic acid and palmitic acid and dietary myristic and palmitic saturated fatty acids and serum palmitic combined with alpha-tocopherol supplementation are associated with increased risk of prostate cancer in a dose-dependent manner. These associations may, however, reflect differences in intake or metabolism of these fatty acids between the precancer cases and controls, rather than being an actual cause.

Prostate cancer is a form of cancer that develops in the prostate, a gland in the male reproductive system. Most prostate cancers are slow growing; however, there are cases of aggressive prostate cancers.

The cancer cells may metastasize (spread) from the prostate to other parts of the body, particularly the bones and lymph nodes. Prostate cancer may cause pain, difficulty in urinating, problems during sexual intercourse, or erectile dysfunction. Other symptoms can potentially develop during later stages of the disease.

Rates of detection of prostate cancers vary widely across the world, with South and East Asia detecting less frequently than in Europe, and especially the United States. Prostate cancer tends to develop in men over the age of fifty. Globally it is the sixth leading cause of cancer-related death in men (in the United States it is the second). Prostate cancer is most common in the developed world with increasing rates in the developing world. However, many men with prostate cancer never have symptoms, undergo no therapy, and eventually die of other unrelated causes. Many factors, including genetics and diet, have been implicated in the development of prostate cancer.

The presence of prostate cancer may be indicated by symptoms, physical examination, prostate-specific antigen

(PSA), or biopsy. Prostate-specific antigen testing increases cancer detection but does not decrease mortality. The United States Preventive Services Task Force in 2012 recommended against screening for prostate cancer using the PSA testing, due to the risk of over-diagnosis and over-treatment with most prostate cancer remaining asymptomatic. The USPSTF concludes that the potential benefit of testing does not outweigh the expected harms. Management strategies for prostate cancer should be guided by the severity of the disease. Many low-risk tumors can be safely followed with active surveillance. Curative treatment generally involves surgery, various forms of radiation therapy, or, less commonly, cryosurgery; hormonal therapy and chemotherapy are generally reserved for cases of advanced disease (although hormonal therapy may be given with radiation in some cases).

The age and underlying health of the man, the extent of metastasis, appearance under the microscope and response of the cancer to initial treatment are important in determining the outcome of the disease. The decision whether or not to treat localized prostate cancer (a tumor that is contained within the prostate) with curative intent is a patient trade-off between the expected beneficial and harmful effects in terms of patient survival and quality of life.

Early prostate cancer usually causes no symptoms. Sometimes, however, prostate cancer does cause symptoms, often similar to those of diseases such as benign prostatic hyperplasia. These include frequent urination, nocturia (increased urination at night), difficulty starting and maintaining a steady stream of urine, hematuria (blood in the urine), and dysuria (painful urination). About a third of patients diagnosed with prostate cancer have one or more such symptoms, while two thirds have no symptoms. Prostate cancer is associated with urinary dysfunction as the prostate gland surrounds the prostatic urethra. Changes within the gland, therefore, directly affect urinary function. Because the *vas deferens* deposits seminal fluid into the prostatic urethra, and secretions from the prostate gland itself are included in semen content, prostate cancer may also cause problems with sexual function and performance, such as difficulty achieving erection or painful ejaculation.

Advanced prostate cancer can spread to other parts of the body, possibly causing additional symptoms. The most common symptom is bone pain, often in the vertebrae (bones of the spine), pelvis, or ribs. Spread of cancer into other bones such as the femur is usually to the proximal part of the bone. Prostate cancer in the spine can also compress the spinal cord, causing leg weakness and urinary and fecal incontinence.

A complete understanding of the causes of prostate cancer remains elusive. The primary risk factors are obesity, age and family history. Prostate cancer is very uncommon in men younger than 45, but becomes more common with advancing age.

The average age at the time of diagnosis is 70. However, many men never know they have prostate cancer. Autopsy studies of Chinese, German, Israeli, Jamaican, Swedish, and Ugandan men who died of other causes have found prostate cancer in thirty percent of men in their 50s, and in eighty percent of men in their 70s. Men who have first-degree family members with prostate cancer appear to have double the risk of getting the disease compared to men without prostate cancer in the family. This risk appears to be greater for men with an affected brother than for men with an affected father. In the United States in 2005, there were an estimated 230,000 new cases of prostate cancer and 30,000 deaths due to prostate cancer. Men with high blood pressure are more likely to develop prostate cancer. There is a small increased risk of prostate cancer associated with lack of exercise. A 2010 study found that prostate basal cells were the most common site of origin for prostate cancers. Genetic background may contribute to prostate cancer risk, as suggested by associations with race, family, and specific gene variants. Men who have a first-degree relative (father or brother) with prostate cancer have twice the risk of developing prostate cancer, and those with two first-degree relatives affected have a fivefold greater risk compared with men with no family history. In the United States, prostate cancer more commonly affects black men than white or Hispanic men, and is also more deadly in black men. In contrast, the incidence and mortality rates for Hispanic men are one third lower than for non-Hispanic whites. Studies of twins in Scandinavia suggest that forty percent of prostate cancer risk can be explained by inherited factors. No single gene is responsible for prostate cancer; many different genes have been implicated. Mutations in *BRCA1* and *BRCA2*, important risk factors for ovarian cancer and breast cancer in women, have also been implicated in prostate cancer. Other linked genes include the Hereditary Prostate cancer gene 1 (HPC1), the androgen receptor, and the vitamin D receptor. TMPRSS2-ETS gene family fusion, specifically TMPRSS2-ERG or TMPRSS2-ETV1/4 promotes cancer cell growth. Loss of cancer suppressor genes, early in the prostatic carcinogenesis, have been localized to chromosomes *8p*, *10q*, *13q*,and *16q*. P53 mutations in the primary prostate cancer are relatively low and are more frequently seen in metastatic settings, hence, p53 mutations are late event in pathology of prostate cancer. Other tumor suppressor genes that are thought to play a role in prostate cancer include PTEN (gene) and KAI1. "Up to 70 percent of men with prostate cancer have lost one copy of the PTEN gene at the time of diagnosis" Relative frequency of loss of E-cadherin and CD44 has also been observed.

While a number of dietary factors have been linked to prostate cancer the evidence is still tentative. Evidence supports little role for dietary fruits and vegetables in prostate cancer occurrence. Red meat and processed meat also appear to have little effect in human studies. Evidence from animals studies however raise concerns. Lower blood levels of vitamin D may increase the risk of developing prostate cancer. This may be linked to lower exposure to ultraviolet (UV) light, since UV light exposure can increase vitamin D in the body.

Green tea may be protective (due to its catechins content), although the most comprehensive clinical study indicates that it has no protective effect. Other holistic methods are also studied. Taking multivitamins more than seven times a week may increase the risks of contracting the disease. This research was unable to highlight the exact vitamins responsible for this increase (almost double), although they suggest that vitamin A, vitamin E and beta-carotene may lie at its heart. It is advised that those taking multivitamins never exceed the stated daily dose on the label. Higher selenium blood levels have been associated with a lower risk of prostate cancer, a trial of supplementation however did not find benefit.

Folic acid supplements have recently been linked to an increase in risk of developing prostate cancer. A ten-year study led by University of Southern California researchers showed that men who took daily folic acid supplements of 1 mg were three times more likely to be diagnosed with prostate cancer than men who took a placebo. High alcohol intake may increase the risk of prostate cancer and interfere with folate metabolism. Low folate intake and high alcohol intake may increase the risk of prostate cancer to a greater extent than the sole effect of either one by itself.

A case control study consisting of 137 veterans addressed this hypothesis and the results were that high folate intake was related to a 79% lower risk of developing prostate cancer and there was no association between alcohol consumption by itself and prostate cancer risk. Folate's effect however was only significant when coupled with low alcohol intake.

There is a significant decrease in risk of prostate cancer with increasing dietary folate intake but this association only remains in individuals with low levels of alcohol consumption. There was no association found in this study between folic acid supplements and risk of prostate cancer.

There are also some links between prostate cancer and medications, medical procedures, and medical conditions. Use of the cholesterol-lowering drugs known as the statins may also decrease prostate cancer risk. Infection or inflammation of the prostate (prostatitis) may increase the chance for prostate cancer while another study shows infection may help prevent prostate cancer by increasing blood to the area. In particular, infection with the sexually transmitted infections chlamydia, gonorrhea, or syphilis seems to increase risk. Finally, obesity and elevated blood levels of testosterone may increase the risk for prostate cancer. There is an association between vasectomy and prostate cancer however more research is needed to determine if this is a causative relationship.

Research released in May 2007, found that US war veterans who had been exposed to Agent Orange had a 48% increased risk of prostate cancer recurrence following surgery.

In 2006, researchers associated a previously unknown retrovirus, Xenotropic MuLV-related virus or XMRV, with human prostate tumors. Subsequent reports on the virus have been contradictory. A group of US researchers found XMRV protein expression in human prostate tumors, while German scientists failed to find XMRV-specific antibodies or XMRV-specific nucleic acid sequences in prostate cancer samples.

The prostate is a part of the male reproductive system that helps make and store seminal fluid. In adult men, a typical prostate is about three centimetres long and weighs about twenty grams. It is located in the pelvis, under the urinary bladder and in front of the rectum.

The prostate surrounds part of the urethra, the tube that carries urine from the bladder during urination and semen during ejaculation. Because of its location, prostate diseases often affect urination, ejaculation, and rarely defecation. The prostate contains many small glands which make about twenty percent of the fluid constituting semen. In prostate cancer, the cells of these prostate glands mutate into cancer cells.

The prostate glands require male hormones, known as androgens, to work properly. Androgens include testosterone, which is made in the testes; dehydroepiandrosterone, made in the adrenal glands; and dihydrotestosterone, which is converted from testosterone within the prostate itself. Androgens are also responsible for secondary sex characteristics such as facial hair and increased muscle mass.

Prostate cancer is classified as an adenocarcinoma, or glandular cancer, that begins when normal semen-secreting prostate gland cells mutate into cancer cells. The region of prostate gland where the adenocarcinoma is most common is the peripheral zone.

Initially, small clumps of cancer cells remain confined to otherwise normal prostate glands, a condition known as carcinoma in situ or prostatic intraepithelial neoplasia (PIN). Although there is no proof that PIN is a cancer precursor, it is closely associated with cancer. Over time, these cancer cells begin to multiply and spread to the surrounding prostate tissue (the stroma) forming a tumor.

Eventually, the tumor may grow large enough to invade nearby organs such as the seminal vesicles or the rectum, or the tumor cells may develop the ability to travel in the bloodstream and lymphatic system. Prostate cancer is considered a malignant tumor because it is a mass of cells that can invade other parts of the body. This invasion of other organs is called metastasis. Prostate cancer most commonly metastasizes to the bones, lymph nodes, and may invade rectum, bladder and lower ureters after local progression. The route of metastasis to bone is thought to be venous as the prostatic venous plexus draining the prostate connects with the vertebral veins.

The prostate is a zinc accumulating, citrate producing organ. The protein ZIP1 is responsible for the active transport of zinc into prostate cells. One of zinc's important roles is to change the metabolism of the cell in order to produce citrate, an important component of semen.

The process of zinc accumulation, alteration of metabolism, and citrate production is energy inefficient, and prostate cells sacrifice enormous amounts of energy (ATP) in order to accomplish this task. Prostate cancer cells are generally devoid of zinc. This allows prostate cancer cells to save energy not making citrate, and utilise the new abundance of energy to grow and spread. The absence of zinc is thought to occur via a silencing of the gene that produces the transporter protein ZIP1. ZIP1 is now called a tumor suppressor gene product for the gene SLC39A1. The cause of the epigenetic silencing is unknown.

Strategies which transport zinc into transformed prostate cells effectively eliminate these cells in animals. Zinc inhibits NF-κB pathways, is anti-proliferative, and induces apoptosis in abnormal cells. Unfortunately, oral ingestion of zinc is ineffective since high concentrations of zinc into prostate cells is not possible without the active transporter, ZIP1.

RUNX2 is a transcription factor that prevents cancer cells from undergoing apoptosis thereby contributing to the development of prostate cancer.

The PI3k/Akt signaling cascade works with the transforming growth factor beta/SMAD signaling cascade to ensure prostate cancer cell survival and protection against apoptosis.

X-linked inhibitor of apoptosis (XIAP) is hypothesized to promote prostate cancer cell survival and growth and is a target of research because if this inhibitor can be shut down then the apoptosis cascade can carry on its function in preventing cancer cell proliferation. Macrophage inhibitory cytokine-1 (MIC-1) stimulates the focal adhesion kinase (FAK) signaling pathway which leads to prostate cancer cell growth and survival.

The androgen receptor helps prostate cancer cells to survive and is a target for many anti cancer research studies; so far, inhibiting the androgen receptor has only proven to be effective in mouse studies. Prostate specific membrane antigen (PSMA) stimulates the development of prostate cancer by increasing folate levels for the cancer cells to use to survive and grow; PSMA increases available folates for use by hydrolyzing glutamated folates. The American Cancer Society's position regarding early detection is "Research has not yet proven that the potential benefits of testing outweigh the harms of testing and treatment. The American Cancer Society believes that men should not be tested without learning about what we know and don't know about the risks and possible benefits of testing and treatment. Starting at age 50, (45 if African American or brother or father suffered from condition before age 65) talk to your doctor about the pros and cons of testing so you can decide if testing is the right choice for you." The only test that can fully confirm the diagnosis of prostate cancer is a biopsy, the removal of small pieces of the prostate for microscopic examination. However, prior to a biopsy, less invasive testing can be conducted.

According to Professor Hardev Pandha, The Prostate Project Chair of Urological Oncology at the University of Surrey's Postgraduate Medical School, a non-invasive test looking for the presence of the protein Engrailed-2 (EN2) in the urine to be more reliable and accurate than existing tests.

"In this study, we showed that the new test was twice as good at finding prostate cancer as the standard PSA test. Only rarely did we find EN2 in the urine of men who were cancer free, so if we find EN2 we can be reasonably sure that a man has prostate cancer. EN2 was not detected in men with non-cancer disorders of the prostate such as prostatitis or benign enlargement. These conditions often cause a high PSA result, causing considerable stress for the patient and sometimes also unnecessary further tests such as prostate biopsies."

There are also several other tests that can be used to gather more information about the prostate and the urinary tract. Digital rectal examination (DRE) may allow a doctor to detect prostate abnormalities. Cystoscopy shows the urinary tract from inside the bladder, using a thin, flexible camera tube inserted down the urethra. Transrectal ultrasonography creates a picture of the prostate using sound waves from a probe in the rectum.

Ultrasound (US) and Magnetic Resonance Imaging (MRI) are the two main imaging methods used for prostate cancer detection. Urologists use transrectal ultrasound during prostate biopsy and can sometimes see a hypoechoic area. But US has poor tissue resolution and thus, is generally not clinically used.

In contrast, prostate MRI has superior soft tissue resolution. MRI is a type of imaging that uses magnetic fields to locate and characterise prostate cancer. Multi-parametric prostate MRI consists of four types of MRI sequences called T2 weighted imaging, T1 weighted imaging, Diffusion Weighted Imaging, MR Spectrocopic Imaging and Dynamic-Contrast Enhanced Imaging.

Genitourinary radiologists use multi-parametric MRI to locate and identify prostate cancer. Currently, MRI is used to identify targets for prostate biopsy using fusion MRI with ultrasound (US) or MRI-guidance alone. In men who are candidates for active surveillance, fusion MR/US guided prostate biopsy detected 33% of cancers compared to 7% with standard ultrasound guided biopsy. Prostate MRI is also used for surgical planning for men undergoing robotic prostatectomy.

It has also shown to help surgeons decide whether to resect or spare the neurovascular bundle, determine return to urinary continence and help assess surgical difficulty. Some prostate advocacy groups believe prostate MRI should be used to screen for prostate cancer—"manogram"— like mammogram is for breast cancer. NIH-funded clinical trials are underway to delineate the value of MRI for some of these applications.

If cancer is suspected, a biopsy is offered expediently. During a biopsy a urologist or radiologist obtains tissue samples from the prostate via the rectum. A biopsy gun inserts and removes special hollow-core needles (usually three to six on each side of the prostate) in less than a second. Prostate biopsies are routinely done on an outpatient basis and rarely require hospitalization. Fifty-five percent of men report discomfort during prostate biopsy.

The tissue samples are then examined under a microscope to determine whether cancer cells are present, and to evaluate the microscopic features (or Gleason score) of any cancer found. Prostate specific membrane antigen is a transmembrane carboxypeptidase and exhibits folate hydrolase activity. This protein is overexpressed in prostate cancer tissues and is associated with a higher Gleason score.

Tissue samples can be stained for the presence of PSA and other tumor markers in order to determine the origin of malignant cells that have metastasized.

Small cell carcinoma is a very rare (1%) type of prostate cancer that cannot be diagnosed using the PSA. As of 2009 researchers are trying to determine the best way to screen for this type of prostate cancer because it is a relatively unknown and rare type of prostate cancer but very serious and quick to spread to other parts of the body.

Possible methods include chromatographic separation methods by mass spectrometry, or protein capturing by immunoassays or immunized antibodies. The test method will involve quantifying the amount of the biomarker PCI, with reference to the Gleason Score. Not only is this test quick, it is also sensitive. It can detect patients in the diagnostic grey zone, particularly those with a serum free to total Prostate Specific Antigen ratio of 10-20%.

The oncoprotein BCL-2, has been associated with the development of androgen-independent prostate cancer due to its high levels of expression in androgen-independent tumours in advanced stages of the pathology. The upregulation of BCL-2 after androgen ablation in prostate carcinoma cell lines and in a castrated-male rat model further established a connection between BCL-2 expression and prostate cancer progression. The expression of Ki-67 by immunohistochemistry may be a significant predictor of patient outcome for men with prostate cancer. ERK5 is a protein that may be used as a marker. ERK5 is present in abnormally high levels of prostate cancer, including invasive cancer which has spread to other parts of the body. It is also present in relapsed cancer following previous hormone therapy. Research shows that reducing the amount of ERK5 found in cancerous cells reduces their invasiveness.

An important part of evaluating prostate cancer is determining the stage, or how far the cancer has spread. Knowing the stage helps define prognosis and is useful when selecting therapies. The most common system is the four-stage TNM system (abbreviated from Tumor/Nodes/Metastases). Its components include the size of the tumor, the number of involved lymph nodes, and the presence of any other metastases.

The most important distinction made by any staging system is whether or not the cancer is still confined to the prostate. In the TNM system, clinical T1 and T2 cancers are found only in the prostate, while T3 and T4 cancers have spread elsewhere. Several tests can be used to look for evidence of spread. These include computed tomography to evaluate spread within the pelvis, bone scans to look for spread to the bones, and endorectal coil magnetic resonance imaging to closely evaluate the prostatic capsule and the seminal vesicles. Bone scans should reveal osteoblastic appearance due to *increased* bone density in the areas of bone metastasis—opposite to what is found in many other cancers that metastasize. After a prostate biopsy, a pathologist looks at the samples under a microscope. If cancer is present, the pathologist reports the grade of the tumor. The grade tells how much the tumor tissue differs from normal prostate tissue and suggests how fast the tumor is likely to grow. The Gleason system is used to grade prostate tumors from 2 to 10, where a Gleason score of 10 indicates the most abnormalities. The pathologist assigns a number from 1 to 5 for the most common pattern observed under the microscope, then does the same for the second-most-common pattern. The sum of these two numbers is the Gleason score. The Prostate cancer screening is an attempt to find unsuspected cancers, and may lead to more invasive follow-up tests such as a biopsy, with cell samples taken for closer study. Options include the digital rectal exam (DRE) and the prostate-specific antigen (PSA) blood test. Such screening is controversial and, in some people, may lead to unnecessary, even harmful, consequences. Routine screening with either a DRE or PSA is not supported by the evidence as there is no mortality benefit from screening. The United States Preventive Services Task Force (USPSTF) recommends against the PSA test for prostate cancer screening in healthy men regardless of age. The USPSTF concludes that the potential benefit of testing does not outweigh the expected harms, "Prostate-specific antigen–based screening results in small or no reduction in prostate cancer–specific mortality and is associated with harms related to subsequent evaluation and treatments, some of which may be unnecessary."

The Centres for Disease Control and Prevention shared that conclusion. The American Society of Clinical Oncology (ASCO) recommends screening be discouraged in those who are expected to live less than ten years, while in those with a life expectancy of greater than ten years a decision should be made by the person in question based on the potential risks and benefits.

There is a significant relation between lifestyle (including food consumption) and cancer prevention.

Two medications which block the conversion of testosterone to dihydrotestosterone, finasteride and dutasteride, have also shown some promise. The use of these medications for primary prevention is still in the testing phase, and they are not widely used for this purpose. A 2008 study found that finasteride reduces the incidence of prostate cancer by 30%, without any increase in the risk of High-Grade prostate cancer. In the original study it turns out that the smaller prostate caused by finasteride means that a doctor is more likely to hit upon cancer nests and more likely to find aggressive-looking cells.

Taking 5-alpha-reductase inhibitors (5-ARIs) can reduce a man's risk of being diagnosed with prostate cancer from around 5–9% to around 4-6% during up to 7 years of treatment.

More frequent ejaculation also may decrease a man's risk of prostate cancer. One study showed that men who ejaculated 3-5 times a week at the age of 15-19 had a decreased rate of prostate cancer when they are old, though other studies have shown no benefit.

The results contradict those of previous studies, that suggested that having many sexual partners, or a high frequency of sexual activity, increases the risk of prostate cancer by up to 40 percent. A key difference may be that these earlier studies defined sexual activity as sexual intercourse, whereas this study focused on the number of ejaculations, whether or not intercourse was involved. An explanation for this discrepancy may be that sex with partners increases the risk of contracting sexually transmitted diseases, including cryptic infections, which increase the risk of cancers. Another study completed in 2004 reported that "Most categories of ejaculation frequency were unrelated to risk of prostate cancer. However, high ejaculation frequency was related to decreased risk of total prostate cancer." The report abstract concluded, "Our results suggest that ejaculation frequency is not related to increased risk of prostate cancer." It is recommended that people maintain a normal weight (limiting consumption of energy dense foods and sugary drinks), eat plant based food, limit red and processed meat, and limit alcohol. Some evidence supports a vegetarian or vegan diet. Consuming fish appears to lower prostate cancer deaths but not the occurrence of prostate cancer. Omega-3 fatty acids are unlikely to prevent prostate cancer. There is no evidence that vitamin supplements affect risk. Trans fats may be associated with an increased risk of cancer but the evidence is still limited.

The first decision to be made in managing prostate cancer is whether any treatment at all is needed. Prostate cancer, especially the most common, low-grade forms found in the typical elderly patient, often grows so slowly that no treatment is required at all. Donald Gleason, the inventor of the Gleason score, advocated for renaming the very common 3+3 prostate "cancer" to prostate adenosis, because he believed it so unlikely to harm the patient. Treatment may also be inappropriate or impossible if the patient has other serious health problems or is not expected to live long enough for symptoms to appear.

Which option is best depends on the stage of the disease, the Gleason score, and the PSA level. Other important factors are age, general health, and patient views about potential treatments and their possible side effects. Because all treatments can have significant side effects, such as erectile dysfunction and urinary incontinence, treatment discussions often focus on balancing the goals of therapy with the risks of lifestyle alterations. A combination of the treatment options is often recommended for managing prostate cancer. The National Comprehensive Cancer Network (NCCN) offers evidence-based guidelines for prostate cancer that can guide treatment choices for specific clinical situations. This requires a good estimation of the patient's long-term health-adjusted life expectancy, because this factor is the most important determinant of survival in newly diagnosed patients. A simplified approach shows how to estimate health-adjusted life expectancy and apply the NCCN guidelines so that patients can have a roadmap to reach the decision recommended for their clinical situation, which they can alter according to their personal values, including fear of cancer and fear of side effects.

Patients can also use a newly developed 18-item questionnaire to learn whether they have good knowledge and understanding about their treatment options before they choose an option. Most newly diagnosed patients who have already made a treatment choice can not correctly answer over half of the questions.

The selection of treatment options involves many factors. For example, if radiation therapy is done first, and fails, then radical prostatectomy is a very technically challenging surgery and may not be feasible. On the other hand, radiation therapy done after surgical failure may have many complications. The desire to maximise subsequent options in case of failure may affect the treatment decision.

Many men diagnosed with low-risk prostate cancer are eligible for *active surveillance*. This term implies careful observation of the tumor over time, with the intention of treatment for cure if there are signs of cancer progression. Active surveillance is *not* synonymous with watchful waiting, an older term which implies no treatment or specific program of monitoring, with the assumption that *palliative*, not curative, treatment would be used if advanced, symptomatic disease develops. Active surveillance involves monitoring the tumor for signs of growth or the appearance of symptoms. The monitoring process may involve serial PSA, physical examination of the prostate, and/or repeated biopsies. The goal of surveillance is to avoid overtreatment and the sometimes serious, permanent side effects of treatment for a slow-growing or self-limited tumor that would never cause any problems for the patient.

This approach is not used for aggressive cancers, but it may cause anxiety for patients who wrongly believe that all cancer is deadly or themselves to have a life-threatening cancer.

For the 50% to 75% of patients with prostate cancer that will cause no harm before the man dies of something unrelated, active surveillance may be the best choice.

Treatment for aggressive prostate cancers involves surgery (i.e. radical prostatectomy), radiation therapy including brachytherapy (prostate brachytherapy) and external beam radiation therapy, High-intensity focused ultrasound (HIFU), chemotherapy, oral chemothcrapeutic drugs (Temozolomide/TMZ), cryosurgery, hormonal therapy, or some combination.

Because of PSA screening, almost 90% of patients are diagnosed when the cancer is localized to the prostate gland and its removal by surgery or radiotherapy will in most cases lead to a cure. Because of this almost 94% of U.S. patients choose treatment. However, in 50% to 75% of these patients the cancer would not have affected their survival even without treatment, and by accepting treatment they have a high chance of sexual, urinary, and bowel side effects. For instance, two-thirds of treated patients cannot get sufficient erections for intercourse, and almost a third have urinary leakage. However, some cancers will grow faster and prostate cancer is the second most common reason of cancer death in U.S. men, after lung cancer.

Although the widespread use of prostate specific antigen (PSA) screening in the USA has resulted in diagnosis at earlier age and cancer stage, the vast majority of cases are still diagnosed in men older than 65 years, and approximately 25% of cases are diagnosed in men older than 75 years.

Though US National Comprehensive Cancer Network guidelines recommend using life expectancy greater than or less than 10 years to help make treatment decisions, in practice, many elderly patients are not offered curative treatment options such as radical prostatectomy (RP) or radiation therapy and are instead treated with hormonal therapy or watchful waiting.

This pattern can be attributed to factors such as medical co-morbidity and patient preferences is regard to quality of life in addition to prostate cancer specific risk factors such as pretreatment PSA, Gleason score and clinical stage. As the average life expectancy increases due to advances in treatment of cardiovascular, pulmonary and other chronic disease, it is likely that more elderly patients will be living long enough to suffer the consequences of their prostate cancer.

Therefore, there is currently much interest in the role of aggressive prostate cancer treatment modalities such as with surgery or radiation in the elderly population who have localized disease.

The results of one randomized controlled trial published by the Scandinavian Prostate Cancer Group 4 evaluated cancer-specific mortality in patients treated with RP compared with watchful waiting.

The patients receiving radical prostatectomy had a relative risk reduction of 30.7% [95% confidence interval 2.5%-50.7%], but an absolute risk reduction of 6% [95% confidence interval 0.5%-11.5%]. The number needed to treat was calculated to be 16. This means that, over the median follow up period of approximately 10 years, 16 patients with localized prostate cancer would need to receive radical prostatectomy rather than watchful waiting in order to prevent one death due to prostate cancer.

Further subset analysis revealed that this benefit did not apply to all ages equally. In men younger than 65 years, patients randomized to receive radical prostatectomy actually had a 10-18% absolute risk reduction in cancer-specific mortality compared to those randomized to watchful waiting.

However, in men older than 65, there was no statistically significant risk reduction even when adjusted for PSA level, Gleason score and tumor stage. Randomized, controlled trials comparing radical prostatectomy, radiation therapy, hormonal therapy and watchful waiting would provide the best evidence for how to best treat elderly patients. If the cancer has spread beyond the prostate, treatment options significantly change, so most doctors that treat prostate cancer use a variety of nomograms to predict the probability of spread. Treatment by watchful waiting/active surveillance, external beam radiation therapy, brachytherapy, cryosurgery, HIFU, and surgery are, in general, offered to men whose cancer remains within the prostate. Hormonal therapy and chemotherapy are often reserved for disease that has spread beyond the prostate. However, there are exceptions: radiation therapy may be used for some advanced tumors, and hormonal therapy is used for some early stage tumors. Cryotherapy (the process of freezing the tumor), hormonal therapy, and chemotherapy may also be offered if initial treatment fails and the cancer progresses.

A vaccine, sipuleucel-T has been found result in a benefit (a four month increase in survival) for men with metastatic prostate cancer.

Most hormone dependent cancers become refractory after one to three years and resume growth despite hormone therapy. Previously considered "hormone-refractory prostate cancer" or "androgen-independent prostate cancer", the term castration-resistant has replaced "hormone refractory" because while they are no longer responsive to castration treatment (reduction of available androgen/testosterone/DHT by chemical or surgical means), these cancers still show reliance upon hormones for androgen receptor activation. Before 2004, all treatments for castration-resistant prostate cancer (CRPC) were considered palliative and not shown to prolong survival. However, there are now several treatments available to treat CRPC that improve survival.

The cancer chemotherapic docetaxel has been used as treatment for (CRPC) with a median survival benefit of 2 to 3 months. Docetaxel's FDA approval in 2004 was significant as it was the first treatment proven to prolong survival in CRPC. In 2010, the FDA approved a second-line chemotherapy treatment known as cabazitaxel.

Off-label use of the oral drug ketoconazole is sometimes used as a way to further manipulate hormones with a therapeutic effect in CRPC. However, many side effects are possible with this drug and abiraterone is likely to supplant usage since it has a similar mechanism of action with less toxic side effects.

A combination of bevacizumab (Avastin), docetaxel, thalidomide and prednisone appears effective in the treatment of CRPC. The immunotherapy treatment with sipuleucel-T is also effective in the treatment of CRPC with a median survival benefit of 4.1 months. The second line hormonal therapy abiraterone (Zytiga) completed a phase 3 trial for CRPC patients who have failed chemotherapy in 2010. Results were positive with overall survival increased by 4.6 months when compared to placebo. On April 28, 2011, the U.S. Food and Drug Administration approved abiraterone acetate in combination with prednisone to treat patients with late-stage (metastatic) castration-resistant prostate cancer who have received prior docetaxel (chemotherapy).

Prognosis

Prostate cancer rates are higher and prognoses are poorer in developed countries than the rest of the world. Many of the risk factors for prostate cancer are more prevalent in the developed world, including longer life expectancy and diets high in red meat. (People who consume larger amounts of meat and dairy also tend to consume fewer portions of fruits and vegetables. It is not currently clear whether both of these factors, or just one of them, contribute to the occurrence of prostate cancer.)

Also, where there is more access to screening programs, there is a higher detection rate. Prostate cancer is the ninth-most-common cancer in the world, but is the number-one non-skin cancer in men from the United States. Prostate cancer affected 18 percent of American men and caused death in three percent in 2005. In Japan, death from prostate cancer was one-fifth to one-half the rates in the United States and Europe in the 1990s.

In India in the 1990s, half of the people with prostate cancer confined to the prostate died within ten years. African-American men have 50–60 times more prostate cancer and prostate cancer deaths than men in Shanghai, China. In Nigeria, two percent of men develop prostate cancer, and 64% of them are dead after two years.

In patients who undergo treatment, the most important clinical prognostic indicators of disease outcome are stage, pre-therapy PSA level, and Gleason score. In general, the higher the grade and the stage, the poorer the prognosis. Nomograms can be used to calculate the estimated risk of the individual patient. The predictions are based on the experience of large groups of patients suffering from cancers at various stages. In 1941, Charles Huggins reported that androgen ablation therapy causes regression of primary and metastatic androgen-dependent prostate cancer. He was awarded the 1966 Nobel Prize for Physiology or Medicine for this discovery. Androgen ablation therapy causes remission in 80-90% of patients undergoing therapy, resulting in a median progression-free survival of 12 to 33 months. After remission, an androgen-independent phenotype typically emerges, wherein the median overall survival is 23–37 months from the time of initiation of androgen ablation therapy. The actual mechanism contributes to the progression of prostate cancer is not clear and may vary between individual patient. A few possible mechanisms have been proposed.

Classification Systems

Many prostate cancers are not destined to be lethal, and most men will ultimately die from causes other than of the disease. Decisions about treatment type and timing may, therefore, be informed by an estimation of the risk that the tumor will ultimately recur after treatment and/or progress to metastases and mortality. Several tools are available to help predict outcomes, such as pathologic stage and recurrence after surgery or radiation therapy. Most combine stage, grade, and PSA level, and some also add the number or percent of biopsy cores positive, age, and/or other information.

- The *D'Amico classification* stratifies men by low, intermediate, or high risk based on stage, grade, and PSA. It is used widely in clinical practice and research settings. The major downside to the 3-level system is that it does not account for multiple adverse parameters (e.g., high Gleason score *and* high PSA) in stratifying patients.
- The Partin tables predict pathologic outcomes (margin status, extraprostatic extension, and seminal vesicle invasion) based on the same three variables and are published as lookup tables.
- The *Kattan nomograms* predict recurrence after surgery and/or radiation therapy, based on data available either at time of diagnosis or after surgery. The nomograms

can be calculated using paper graphs or software available on a website or for handheld computers. The Kattan score represents the likelihood of remaining free of disease at a given time interval following treatment.

- The UCSF *Cancer of the Prostate Risk Assessment (CAPRA) score* predicts both pathologic status and recurrence after surgery. It offers comparable accuracy as the Kattan preoperative nomogram, and can be calculated without paper tables or a calculator. Points are assigned based on PSA, Grade, stage, age, and percent of cores positive; the sum yields a 0–10 score, with every 2 points representing roughly a doubling of risk of recurrence. The CAPRA score was derived from community-based data in the CaPSURE database. It has been validated among over 10,000 prostatectomy patients, including patients from CaPSURE; the SEARCH registry, representing data from several Veterans Administration and active military medical centres; a multi-institutional cohort in Germany; and the prostatectomy cohort at Johns Hopkins University. More recently, it has been shown to predict metastasis and mortality following prostatectomy, radiation therapy, watchful waiting, or androgen deprivation therapy.

As of 2011, prostate cancer is the second most frequently diagnosed cancer and the sixth leading cause of cancer death in males worldwide. Rates of prostate cancer vary widely across the world. Although the rates vary widely between countries, it is least common in South and East Asia, more common in Europe, and most common in the United States. According to the American Cancer Society, prostate cancer is least common among Asian men and most common among black men, with figures for white men in between. The average annual incidence rate of prostate cancer between 1988 and 1992 among Chinese men in the United States was 15 times higher than that of their counterparts living in Shanghai and Tianjin. However, these high rates may be affected by increasing rates of detection. Many suggest that prostate cancer may be under reported, yet BPH incidence in China and Japan is similar to rates in Western countries., Prostate cancer develops primarily in men over fifty. It is the most common type of cancer in men in the United States, with 186,000 new cases in 2008 and 28,600 deaths. It is the second leading cause of cancer death in U.S. men after lung cancer.

In the United Kingdom it is also the second most common cause of cancer death after lung cancer, where around 35,000 cases are diagnosed every year and of which around 10,000 die of it. Many factors, including genetics and diet, have been implicated in the development of prostate cancer. The Prostate Cancer Prevention Trial found that finasteride reduces the incidence of prostate cancer by 30%. There had been a controversy about this also increasing the risk of more aggressive cancers, but more recent research showed this may not be the case. More than 80% of men will develop prostate cancer by the age of 80. However, in the majority of cases, it will be slow-growing and harmless. In such men, diagnosing prostate cancer is overdiagnosis—the needless identification of a technically aberrant condition that will never harm the patient—and treatment in such men exposes them to all of the adverse effects, with no possibility of extending their lives. Although the prostate was first described by Venetian anatomist Niccolò Massa in 1536, and illustrated by Flemish anatomist Andreas Vesalius in 1538, prostate cancer was not identified until 1853.

Prostate cancer was initially considered a rare disease, probably because of shorter life expectancies and poorer detection methods in the 19th century. The first treatments of prostate cancer were surgeries to relieve urinary obstruction. Removal of the entire gland (radical perineal prostatectomy) was first performed in 1904 by Hugh H. Young at Johns Hopkins Hospital. Surgical removal of the testes (orchiectomy) to treat prostate cancer was first performed in the 1890s, but with limited success. Transurethral resection of the prostate (TURP) replaced radical prostatectomy for symptomatic relief of obstruction in the middle of the 20th century because it could better preserve penile erectile function. Radical retropubic prostatectomy was developed in 1983 by Patrick Walsh. This surgical approach allowed for removal of the prostate and lymph nodes with maintenance of penile function. In 1941, Charles B. Huggins published studies in which he used estrogen to oppose testosterone production in men with metastatic prostate cancer. This discovery of "chemical castration" won Huggins the 1966 Nobel Prize in Physiology or Medicine. The role of the gonadotropin-releasing hormone (GnRH) in reproduction was determined by Andrzej W. Schally and Roger Guillemin, who both won the 1977 Nobel Prize in Physiology or Medicine for this work. GnRH receptor agonists, such as leuprolide and goserelin, were subsequently developed and used to treat prostate cancer. Radiation therapy for prostate cancer was first developed in the early 20th century and initially consisted of intraprostatic radium implants. External beam radiotherapy became more popular as stronger [X-ray] radiation sources became available in the middle of the 20th century. Brachytherapy with implanted seeds (for prostate cancer) was first described in 1983. Systemic chemotherapy for prostate cancer was first studied in the 1970s. The initial regimen of cyclophosphamide and 5-fluorouracil was quickly joined by multiple regimens using a host of other systemic chemotherapy drugs.

Cell-of-Origin

A series of studies published in *Science* involved introduced viruses known to cause cancerous mutation in prostate cells: AKT, ERG, and AR into isolated samples of basal and luminal cells and grafted the treated tissue into mice. After 16 weeks, none of the luminal samples had undergone malignant mutation, while the basal samples had mutated into prostate-like tubules which had then developed malignancy and formed cancerous tumors, which appeared identical to human samples under magnification. This led to the conclusion that the prostate basal cell may be the most likely "site of origin" of prostate cancer.

Society and Culture

People with prostate cancer generally encounter significant disparities in awareness, funding, media coverage, and research—and therefore, inferior treatment and poorer outcomes—compared to other cancers of equal prevalence. In 2001, *The Guardian* noted that Britain had 3,000 nurses specialising in breast cancer, compared to only one for prostate cancer. It also discovered that the waiting time between referral and diagnosis was two weeks for breast cancer but three months for prostate cancer. A 2007 report by the U.S.-based National Prostate Cancer Coalition stated that for every prostate cancer drug on the market, there were seven used to treat breast cancer.

The Times also noted an "anti-male bias in cancer funding" with a four to one discrepancy in the United Kingdom by both the government and by cancer charities such as Cancer Research UK. Equality campaigners such as author Warren Farrell cite such stark spending inequalities as a clear example of governments unfairly favouring women's health over men's health. Disparities also extend into areas such as detection, with governments failing to fund or mandate prostate cancer screening while fully supporting breast cancer programs. For example, a 2007 report found 49 U.S. states mandate insurance coverage for routine breast cancer screening, compared to 28 for prostate cancer. Prostate cancer also experiences significantly less media coverage than other, equally prevalent cancers, with a study by Prostate Coalition showing 2.6 breast cancer stories for each one covering cancer of the prostate.

Prostate Cancer Awareness Month takes place in September in a number of countries. A light blue ribbon is used to promote the cause. A genistein derivative KBU2046 is under investigation for prostate cancer. MDV3100 is in phase III trials for HRPC (chemo-naive and post-chemo patient populations). Alpharadin completed a phase 3 trial for CRPC patients with bone metastasis. A pre-planned interim analysis showed improved survival and quality of life. The study was stopped for ethical reasons to give the placebo group the same treatment. Apharadin uses bone targeted Radium-223 isotopes to kill cancer cells by alpha radiation. Alpharadin is an investigational agent and is not approved for marketing by the European Medicines Agency (EMA), the U.S. Food and Drug Administration (FDA), or any other health authorities. Scientists have established a few prostate cancer cell lines to investigate the mechanism involved in the progression of prostate cancer. LNCaP, PC-3 (PC3), and DU-145 (DU145) are commonly used prostate cancer cell lines. The LNCaP cancer cell line was established from a human lymph node metastatic lesion of prostatic adenocarcinoma. PC-3 and DU-145 cells were established from human prostatic adenocarcinoma metastatic to bone and to brain, respectively. LNCaP cells express androgen receptor (AR); however, PC-3 and DU-145 cells express very little or no AR. AR, an androgen-activated transcription factor, belongs to the steroid nuclear receptor family. Development of the prostate is dependent on androgen signaling mediated through AR, and AR is also important during the development of prostate cancer. The proliferation of LNCaP cells is androgen-dependent but the proliferation of PC-3 and DU-145 cells is androgen-insensitive.

Elevation of AR expression is often observed in advanced prostate tumors in patients. Some androgen-independent LNCaP sublines have been developed from the ATCC androgen-dependent LNCaP cells after androgen deprivation for study of prostate cancer progression. These androgen-independent LNCaP cells have elevated AR expression and express prostate specific antigen upon androgen treatment. The paradox is that androgens inhibit the proliferation of these androgen-independent prostate cancer cells. At present, an active area of research and non-clinically applied investigations involve non-invasive methods of prostate tumor detection. Adenoviruses modified to transfect tumor cells with harmless yet distinct genes (such as luciferase) have proven capable of early detection. So far, however, this area of research has been tested only in animal and LNCaP cell models.

EN2: Presence of the EN2 (gene) in urine has been correlated to a high probability of prostate cancer.

PCA3: Another potential non-invasive method of early prostate tumor detection is through a molecular test that detects the presence of cell-associated PCA3 mRNA in fluid massaged from the prostate by the doctor and first-void urinated out within a limited amount of urine into the specimen container. PCA3 mRNA is expressed almost exclusively by prostate cells and has been shown to be highly over-expressed in prostate cancer cells. The test result is currently reported as a specimen ratio of PCA3 mRNA to PSA mRNA. Although not a replacement for serum PSA level, the PCA3 test is an additional tool to help decide whether, in men suspected of having prostate cancer (especially if an initial biopsy fails to explain the elevated serum PSA), a biopsy/rebiopsy is really needed. The higher the expression of PCA3 in the sample, the greater the likelihood of a positive biopsy; i.e., the presence of cancer cells in the prostate.

Early Prostate Cancer Antigen-2

A new blood test for early prostate cancer antigen-2 (EPCA-2) may alert men if they have prostate cancer and how aggressive it will be. Thrombophlebitis is associated with an increased risk of prostate cancer and may be a good way for physicians to remind themselves to screen patients with thrombophlebitis for prostate cancer as well since these two are closely linked.

Prostasomes

Epithelial cells of the prostate secrete prostasomes as well as PSA. Prostasomes are membrane–surrounded, prostate-derived organelles that appear extracellularly, and one of their physiological functions is to protect the sperm from attacks by the female immune system. Cancerous prostate cells continue to synthesize and secrete prostasomes, and may be shielded against immunological attacks by these prostasomes. Research of several aspects of prostasomal involvement in prostate cancer has been performed.

Bones

Mounting evidence indicates that the amount and type of fat in the diet can have important effects on bone health. Most of this evidence is derived from animal studies. The data from one study indicated that bone mineral density is negatively associated with saturated fat intake, and that men may be particularly vulnerable.

Unsaturated Fat

An unsaturated fat is a fat or fatty acid in which there is at least one double bond within the fatty acid chain. A fat molecule is monounsaturated if it contains one double bond, and polyunsaturated if it contains more than one double bond. Where double bonds are formed, hydrogen atoms are eliminated. Thus, a saturated fat has no double bonds, has the maximum number of hydrogens bonded to the carbons, and therefore is “saturated” with hydrogen atoms. In cellular metabolism, unsaturated fat molecules contain somewhat less energy (i.e., fewer calories) than an equivalent amount of saturated fat. The greater the degree of unsaturation in a fatty acid (i.e., the more double bonds in the fatty acid) the more vulnerable it is to lipid peroxidation (rancidity). Antioxidants can protect unsaturated fat from lipid peroxidation.

Double bonds may be in either a *cis* or a *trans* isomer, depending on the geometry of the double bond. In the *cis* conformation, hydrogen atoms are on the same side of the double bond, whereas, in the *trans* conformation, they are on opposite sides. Saturated fats are popular with manufacturers of processed foods because they are less vulnerable to rancidity and are, in general, more solid at room temperature than *unsaturated* fats. Unsaturated chains have a lower melting point, hence increasing fluidity of the cell membranes.

Although both monounsaturated and polyunsaturated fats can replace saturated fat in the diet, trans unsaturated fats should be avoided. Replacing saturated fats with unsaturated fats helps to lower levels of total cholesterol and LDL cholesterol in the blood. Trans unsaturated fats are particularly risky because the double bond stereochemistry allows the fat molecules to assume a linear conformation, which leads to efficient packing (i.e., plaque formation). The geometry of the cis double bond introduces a bend in the molecule, thereby precluding stable formations. Natural sources of fatty acids are rich in the cis isomer.

Although polyunsaturated fats are protective against cardiac arrhythmias, a study of post-menopauseal women with a relatively low fat intake showed that polyunsaturated fat is positively associated with progression of coronary atherosclerosis, whereas monounsaturated fat is not. This probably is an indication of the greater vulnerability of polyunsaturated fats to lipid peroxidation, against which vitamin E has been shown to be protective.

Examples of unsaturated fats are palmitoleic acid, oleic acid, myristoleic acid, linoleic acid, and arachidonic acid. Foods containing unsaturated fats include avocado, nuts, and vegetable oils such as canola and olive oils. Meat products contain both saturated and unsaturated fats.

Although unsaturated fats are conventionally regarded as 'healthier' than saturated fats, the United States Food and

Drug Administration (FDA) recommendation stated that the amount of unsaturated fat consumed should not exceed 30% of one's daily caloric intake (or 67 grams given a 2000 Calorie diet). The new dietary guidelines have eliminated this recommendation at the request of the meat and dairy industries. Most foods contain both unsaturated and saturated fats. Marketers advertise only one or the other, depending on which one makes up the majority. Thus, various unsaturated fat vegetable oils, such as olive oils, also contain saturated fat).

Role of Dietary Fats in Insulin Resistance

Incidence of Insulin resistance is lowered with diets higher in monounsaturated fats (especially oleic acid), while the opposite is true for diets high in polyunsaturated fats (especially large amounts of arachidonic acid) as well as saturated fats (such as arachidic acid), these ratios can be indexed in the phospholipids of human skeletal muscle and in other issues as well. This relationship between dietary fats and insulin resistance is presumed secondary to the relationship between insulin resistance and inflammation, which is partially modulated by dietary fat ratios (Omega3/6/9) with both omega 3 and 9 thought to be anti-inflammatory, and omega 6 pro-inflammatory (as well as by numerous other dietary components, particularly polyphenols, and by exercise as well, with both of these anti-inflammatory). Although both pro-inflammatory and anti-inflammatory types of fat are biologically necessary, fat dietary ratios in most US diets are skewed towards Omega 6, with subsequent disinhibition of inflammation and potentiation of insulin resistance.

Membrane Composition as a Metabolic Enabler

Cell membranes of mammals have a higher composition of polyunsaturated fat (DHA, omega-3 fatty acid) and a lower composition of monounsaturated fat than do reptiles. Higher polyunsaturated membrane content gives greater membrane fluidity (and functionality), commensurate with the higher metabolic rate of the warm-blooded species. In fish, however, increasingly cold environments lead to increasingly high cell membrane content of both monounsaturated and polyunsaturated fatty acids, to maintain greater membrane fluidity (and functionality) at the lower temperatures.

Chapter 6

Soft Organic Biomolecules

Drug research and design has made significant advancements within the past few decades, and still remains an industry in progress. With many different sub-categories in the world of drug design, one particular branch has received a growing amount of attention: creating disease-specific drug delivery vehicles.

Drug delivery devices would be predominantly used to treat diseases such as cancer, because current treatments (chemotherapy) also cause harm to essential and non-pathogenic bodily tissues. Soft, amphiphilic polymers may hold much utility in creating efficient drug delivery vehicles. The desired characteristics of the polymers are as follows: biodegradable, biocompatible, and high molecular weight.

Being so, the widely abundant Poly Lactic Acid polymer serves a viable foundation for modelling the polymers after. In utilising several known synthetic methodologies, researchers have proposed a way in creating potentially novel amphiphilic biomolecules. Reaction of an aldehyde with a specifically designed isocyanide yields the necessary α-hydroxy acid monomer needed to create a Poly(α-hydroxy acid) polymer. Since any aldehyde can be used, a lot of functionality can be incorporated into the polymer backbone. In particular, a norbornene moiety can be used. Norbornene can participate in a widely used Ring Opening Metathesis Polymerization (ROMP). Using two modes of polymerization in tandem opens up a vast amount of functionalization and drug incorporation, as well as fine-tuned adjustments on amphiphilicity. The picture at the right shows a synthetic scheme in obtaining the bifunctionalized α-hydroxy acid monomer. Amphiphilic polymers will create micelle aggregates with hydrophilic components exposed in the body's aqueous environment. Poly Ethylene Glycol can be used for the amphiphilic zone, which will allow the drug-encapsulated aggregates to dissolve and be compatible with a patient's biochemistry. With disease specific anibodies incorporated into the polymer's functionality, the micelle aggregates can seek out the cancer, and inflict a lethal dose of medicine.

RNA Synthesis

A useful application of RNA analysis can allow scientists to synthesis specific RNA sequences for intended functions. The process itself is rather simple. First, a very large amount of completely random RNA sequences are synthesized with really no thought to actual use.

Then the collection of RNA are eluted down a column that specifically selects for RNA that fits a desired function such as ATP-binding where the column will just be filled with ATP bonded molecules.

Naturally, only the RNA that bind to the function will stay in the column and those can be isolated for further testing by just eluting the rest of the column. Through this random selection process, scientists can pretty much create RNA to conduct any function. The following synthesis can be summarised in a few steps:

1. Initially, a randomized RNA pool containing of any variety molecules is present in a given population inserted into an ATP affinity column.
2. A selection process occurs in which only specific molecules are isolated with desired binding or reactivity properties.
3. The remaining RNA population that have remained the isolation
   process are then intensely amplified through Polymerase Chain Reaction (PCR).
4. Errors that occur within the course of replication are seen as additional variations introduced into this generation of RNA. Errors are induced through reverse transcription of RNA to DNA and back to RNA to increase

the likelihood of error and mutation. The new population would consequently be introduced back into the column for further ATP analysis. Eventually structures that emerge out of the tube are assumed to be plausible structures of what the RNA could have existed in the past.

Amplified Ancient DNA

The use of PCR to amplify DNA has been used to sequence the DNA of ancient organisms, such as the Neanderthal. Results showed that the DNA of homo sapiens have between 22 and 36 substitutions, as opposed to 55 from a chimpanzee. These results showed that the Neanderthal and homo sapiens shared a common ancestor.

Types of Libraries

The first combinatorial libraries where based on peptides (consisting of few of their monomeric units, the 20 animo acids). Other oligomeric libraries (from carbohydrates, nucleic acids, etc.) were also made out of adding component A to B, resulting in AB, then adding C to obtain ABC, adding D...

This was the main technique for creating combinatorial libraries until the synthesis of benzodiazepine (fusion of benzene ring with diazepine ring) libraries was performed by DeWitt[3]. Instead of using the same technique as the oligomeric libraries, these libraries were based on the attachment of several R substituent chains to a molecule in different positions.

The number of products obtained in the library is calculated d by multiplying the number of substituent chains used on the main molecule, in this case benzodiazepine. The R groups are referred to as functionalities, while the points where they attach in the central molecule are called points of diversity.

Library Synthesis

There are two ways of synthesizing libraries through combinatorial chemistry: 1. Using solid-phase synthesis 2. Through solution-phase synthesis

Solid-phase synthesis This technique allows an easy isolation and manipulation of the product, as well as having simple control systems (automation). Excess reagents can be used to complete reactions. Still it has certain disadvantages, such as the specificity of the solid support that don't allow certain chemical reactions to be performed. Solution-phase synthesis On the contrary, solution-phase synthesis allows a better manipulation of the sample (once isolated) in any amounts without being concerned about the solid support. Its disadvantage is difficult sample isolation, reason why scientists are working on implementing the use of reactants that give pure products (or at least products that can be easily purified).

Diversity-Oriented Libraries

Even though combinatorial chemistry has been an essential part of early drug discovery for more than two decades, so far only one combinatorial chemistry-synthesized chemical has been approved for clinical use by FDA (sorafenib, a multikinase inhibitor used for renal cancer). The analysis of poor success rate of the approach has been suggested to connect with the limited chemical space covered by products of combinatorial chemistry. When comparing the properties of compounds in combinatorial chemistry libraries to those of approved drugs and natural products, it was noted that combinatorial chemistry libraries suffer particularly from the lack of chirality, as well as structure rigidity, which are both widely regarded as drug-like properties. Even though natural product drug discovery has not been the most fashionable trend in pharmaceutical industry in recent times, a large proportion of new chemical entities still is nature-derived compounds, so it has been suggested that effectiveness of combinatorial chemistry could be improved by enhancing the chemical diversity of screening libraries. As chirality and rigidity are the two most important features distinguishing approved drugs and natural products from compounds in combinatorial chemistry libraries, these are the two issues emphasized in so-called diversity oriented libraries, for example compound collections that aim at coverage of the chemical space, instead of just huge number of compounds.

Library Management

As we all know, integrating all types of chemistry, biology, and mixed information is exceptionary difficult and important combinatorial chemistry techniques. Thus, without managing all the information, library can not be function as one expected.

For instance, if one scientist are trying to synthesize product "A," library must be able to show all the information necessary to synthesize the "A" such as, key reagent, cost, or if the reaction procedures are reliable etc... it is because the question that one will ask is different depends on the researchers.

Molecular Evolution

Evolution requires three processes:

1. The generation of a diverse population
2. The selection of members based on some criterion of fitness
3. Reproduction to enrich the population in these more-fit members.

Molecular evolution primarily occurs due to mutations, which are changes in the genetic material of a cell.

They can occur during copying errors or due to exposure to chemicals and radiation.

Natural selection removes the less favourable mutations. Key topics under molecular evolution are the study of evolution of enzyme function and the study of nucleic acids. Nucleic acids have shown the divergence of species because they are denoted as the "molecular clock."

Proteins in Industry

Protein adsorption refers to the adhesion of proteins to solid surfaces. This phenomenon is an important issue in the food processing industry, particularly in milk processing and wine and beer making.

Excessive adsorption, or protein fouling, can lead to health and sanitation issues, as the adsorbed protein is very difficult to clean and can harbor bacteria, as is the case in biofilms. Product quality can be adversely affected if the adsorbed material interferes with processing steps, like pasteurization. However, in some cases protein adsorption is used to improve food quality, as is the case in fining of wines.

Protein Adsorption

Protein adsorption and protein fouling can cause major problems in the food industry (particularly the dairy industry) when proteins from food adsorb to processing surfaces, such as stainless steel or plastic (e.g. polypropylene). Protein fouling is the gathering of protein aggregates on a surface. This is most common in heating processes that create a temperature gradient between the equipment and the bulk substance being heated.

In protein-fouled heating equipment, adsorbed proteins can create an insulating layer between the heater and the bulk material, reducing heating efficiency. This leads to inefficient sterilization and pasteurization. Also, proteins stuck to the heater may cause a burned taste or colour in the bulk material. Additionally, in processes that employ filtration, protein aggregates that gather on the surface of the filter can block the flow of the bulk material and greatly reduce filter efficiency.

Examples of Adsorption

Beer Stone

Beer stone is a buildup that forms when oxalate, proteins, and calcium or magnesium salts from the grains and water in the beer brewing process precipitate and form scale on kegs, barrels and tap lines. The minerals adsorb to the surface of the container first, driven by charge attractions. Proteins are often coordinated to these minerals in the solution and can bind with them to the surface. In other cases proteins also adsorb to the minerals on the surface, making deposits difficult to remove, as well as providing a surface that can easily harbor microorganisms. If built-up beer stone inside tap lines flakes off, it can negatively affect the quality of the finished product by making beer hazy and contributing "off" flavours. It is also harmful from a nutritional standpoint: oxalates can decrease absorption of calcium in the body, in addition to increasing risk of kidney stone formation.

Wine Making

Grape and wine proteins tend to aggregate and form hazes and sediment in finished wines, especially white wines. Haze-causing proteins can persist in wine due to low settling velocities or charge repulsion on individual particles. Fining agents, such as bentonite clays, are used to clarify wine by removing these proteins. Also, proteinaceous agents such as albumin, casein, or gelatin are used in wine clarification.

Biofilms

A biofilm is a community of microorganisms adsorbed to a surface. Microorganisms in biofilms are enclosed in a polymeric matrix consisting of exopolysaccharides, extracellular DNA and proteins. Seconds after a surface (usually metal) is placed in a solution, inorganic and organic molecules adsorb onto the surface. These molecules are attracted mainly by Coulombic forces, and can adhere very strongly to the surface.

This first layer is called the conditioning layer, and is necessary for the microorganisms to bind to the surface. These microorganisms then attach reversibly by Van der Waals forces, followed by irreversible adhesion through self-produced attachment structures such as pili or flagella. Biofilms form on solid substrates such as stainless steel. A biofilm's enclosing polymeric matrix offers protection to its microbes, increasing their resistance to detergents and cleaning agents. Biofilms on food processing surfaces can be a biological hazard to food safety. Increased chemical resistance in biofilms can lead to a persistent contamination condition.

Dairy Industry

Thermal treatment of milk by indirect heating (e.g. pasteurization) to reduce microbial load and increase shelf life is generally performed by a plate heat exchanger.

Heat exchanger surfaces can become fouled by adsorbed milk protein deposits. Fouling is initiated by formation of a protein monolayer at room temperature, followed by heat induced aggregation and deposition of whey protein and calcium phosphate deposits. Adsorbed proteins decrease efficiency of heat transfer and potentially affect product quality by preventing adequate heating of milk.

Mechanisms for Protein Adsorption

The common trend in all examples of protein adsorption in the food industry is that of adsorption to minerals adsorbed to the surface first. This phenomenon has been studied but it is not well understood. Spectroscopy of proteins adsorbed onto clay-like minerals show variations in the C=O and N-H bond stretches, meaning that these bonds are involved in the protein binding.

Biochemical

While the aggregates can explain much of the protein fouling found in milk processing, this does not account for it all. A third type of fouling has been discovered that is explained by the chemical interactions of the denatured β-lg proteins.

β-lg contains 5 cysteine residues, four of which are covalently bonded to each other, forming an S-S bond. When β-lg is denatured, the fifth cysteine residue is exposed to the water. This residue then bonds to other β-lg proteins, including those already adsorbed to the surface. This produces a strong interaction between the denatured proteins and the surface of the container.

Isotherms

Isotherms are used to quantify the amount of adsorbed protein on a surface at a constant temperature, depending on the concentration of protein above the surface. Researchers have used a Langmuir-type isotherm model to describe experimental values for protein adsorption.

In this equation

- is the amount of adsorbed protein
- is the surface area per molecule
- is the partial molar volume of protein
- is the negative of the Gibbs Free Energy of adsorption per unit area and
- is the equilibrium protein concentration.

This equation has been applied to a laboratory setting of protein adsorption at temperatures higher than 50°C from a model solution of protein and water. It is especially useful for modelling protein fouling in milk processing.

Removal of Adsorbed Proteins

Adsorbed proteins are among the most difficult food soils to remove from food contact surfaces. In particular, heat-denatured proteins (such as those found in dairy industry applications) adhere tightly to surfaces and require strong alkaline cleaners for removal. It is important that cleaning methods are capable of removing both visible and non-visible protein soils. Nutrients for bacterial growth must be removed as well as biofilms that may have built up on the food contact surface.

Proteins are water-insoluble, slightly soluble in acidic solutions and soluble in alkaline solutions, which limits the type of cleaner that can be used to remove protein from the surface. Generally speaking, highly alkaline cleaners with peptizing and wetting agents are most effective in protein removal on food contact surfaces. Cleaning temperature is also a concern for effective protein removal. As temperature increases, the activity of the cleaning compound increases, making soil removal easier. However, at higher temperatures (> 55°C) proteins denature and cleaning efficacy is reduced.

Alkaline Cleaners

Alkaline cleaners are classified as compounds with pH 7-14. Proteins are most effectively removed from surfaces by cleaners with a pH of 11 or higher. An example of a strong alkaline cleaning agent is sodium hydroxide, also called caustic soda.

Although sodium hydroxide (NaOH) can cause corrosion on food contact surfaces such as stainless steel, it is the preferred cleaning agent for protein removal due to its efficacy in dissolving proteins and dispersing/emulsifying food soils.

Silicates are often added to these cleaners to reduce corrosion on metal surfaces. The mechanism of alkaline cleaning action in proteins follows a three-step process:

1. Gel formation: Upon contact with the alkaline solution, the protein soil swells and forms a removable gel.
2. Protein removal: The protein gel is removed through mass transfer, while the cleaning agent continues to diffuse through the soil, increasing gel formation.
3. Decay stage: The protein gel has been eroded to the point where it is a thin deposit. Removal at this stage is governed by shear stress forces and mass transfer of the gel.

Hypochlorite is often added to alkaline cleaners to peptize proteins. Chlorinated cleansers work by oxidizing sulfide crosslinks in proteins. Cleaning speed and efficiency is improved due to increased diffusion of the cleaner into the soil matrix, now composed of smaller, more soluble proteins.

Enzyme Cleaners

Enzyme-based cleaners are especially useful for biofilm removal. Bacteria are somewhat difficult to remove with traditional alkaline or acid cleaners. Enzyme cleaners are more effective on biofilms since they work as proteases by breaking down proteins at bacterial attachment sites.

They work at maximum efficiency at high pH and at temperatures below 60°C. Enzyme cleaners are an increasingly attractive alternative to traditional chemical cleaners because of biodegradability and other environmental factors, such as reduced wastewater generation and energy savings from using cold water. However, they are typically more expensive than alkaline or acid cleaners.

Proteus Industries

Proteus Industries is a food ingredient manufacturer specialising in patented, all-natural muscle protein which is used to reduce fat, calories and salt, while increasing moisture, in fried or par-fried products.

A product of more than 15 years of research, Proteus Industries was founded in 2001 after Stephen D. Kelleher, Ph.D, founder and chief scientist, discovered a process to recover 95% of pure and functional protein from animal muscle that can be used in a variety of ways, including: reduced fat absorption, improved moisture retention, extended shelf life, pathogen control, substitute for high-fat ingredients such as butter oil, flavour carrier and increased protein content. The company is unrelated to Proteus Industries Inc., a manufacturer of liquid flow switches and metres founded in 1978. Stephen D. Kelleher, Ph.D. is an accomplished scientist and food chemist. Following many years of protein and lipid research, Dr. Kelleher founded Proteus Industries in 2001 and he currently serves as Chief Scientist.

Prior to founding Proteus, Dr. Kelleher co-founded Advanced Protein Technologies, where he remains a director. He spent many years as a researcher at the UMass-Amherst Marine Station in Gloucester, MA. Dr. Kelleher has also worked in a variety of food industry roles with companies such as Campbell Soup Company.

Dr. Kelleher is a graduate of the University of Massachusetts Amherst, from which he received his Ph.D. in Food chemistry. He also earned an M.S. from Cornell University. He has received numerous honours and awards and maintains membership in organisations including American Chemical Society and the Institute of Food Technologists.

Technology

The proteins, recovered by isolating proteins from muscle tissue, are used to improve food quality in a variety of different ways. They are all-natural and not genetically modified. They also meet GRAS Generally recognised as safe standards established by the Food and Drug Administration Food and Drug Administration. The process can be replicated across species and can be applied to fish, chicken, beef, pork and other animal muscle and the resulting isolates can be used in liquid and powder forms in order to achieve the desired benefits in food processing and cooking.

Protein Expression

Protein expression is a subcomponent of gene expression. It consists of the stages after DNA has been transcribed to messenger RNA (mRNA). The mRNA is then translated into polypeptide chains, which are ultimately folded into proteins. Protein expression is commonly used by proteomics researchers to denote the measurement of the presence and abundance of one or more proteins in a particular cell or tissue.

Protein expression systems are very widely used in the life sciences, biotechnology and medicine. Molecular biology research uses numerous proteins and enzymes many of which are from expression systems; particularly DNA polymerase for PCR, reverse transcriptase for RNA analysis and restriction endonucleases for cloning. There are also significant medical applications for expression systems, notably the production of human insulin to treat diabetes. Protein expression systems are used to produce certain proteins in biotechnology and industry, and more recently to produce sets (combinatorial series) of protein that are screened for drug discovery purpose.

Expression Systems

Commonly used protein expression systems include those derived from bacteria, yeast, baculovirus/insect, and mammalian cells.

Cell-Based Systems

The oldest and most widely used expression systems are cell-based and may be defined as the "*combination of an expression vector, its cloned DNA, and the host for the vector that provide a context to allow foreign gene function in a host cell, that is, produce proteins at a high level*". Expression is often done to a very high level and therefore referred to as overexpression. There are many ways to introduce foreign DNA to a cell for expression, and there are many different host cells which may be used for expression - each expression system has distinct advantages and liabilities. Expression systems are normally referred to by the host and the DNA source or the delivery mechanism for the genetic material.

For example, common hosts are bacteria (such as *E.coli*, *B. subtilis*), yeast (such as *S.cerevisiae*) or eukaryotic cell lines. Common DNA sources and delivery mechanisms are viruses (such as baculovirus, retrovirus, adenovirus), plasmids, artificial chromosomes and bacteriophage (such as lambda). The best expression system of choice depends on the gene involved, for example the *Saccharomyces cerevisiae* is often preferred for proteins that require significant post-translational modification and Insect or mammal cell lines are used when human-like splicing of the mRNA is required. Nonetheless, bacterial expression has the advantage of easily producing large amounts of protein, which is required for X-ray crystallography or nuclear magnetic resonance experiments for structure determination.

Because bacteria are prokaryotes, they are not equipped with the full enzymatic machinery to accomplish the required post-translational modifications or molecular folding. Hence, multi-domain eukaryotic proteins expressed in bacteria often are non-functional. Also, many proteins become insoluble as inclusion bodies that are very difficult to recover without harsh denaturants and subsequent cumbersome protein-refolding procedures. To address theses concerns, expressions systems using several eukaryotic cells were developed for applications requiring the proteins be conformed as in, or closer to superior organisms: cells of plants (i.e. tobacco), of insects or mammalians (i.e. bovines) are transfected with genes and cultured in suspension and even as tissues or whole organisms, to produce fully folded proteins. Mammalian in vivo expression systems have however low yield and other eventual limitations (time-consuming, toxicity to host cells,..). To combine the high yield/productivity and scalable protein features of bacteria and yeast, and advanced epigenetic features of plants, insects and mammalians systems, other protein expression systems are developed using unicellular eukaryotes (i.e. non-pathogenic 'Leishamania' cells).

Bacteria Systems

E. coli is one of the most widely used expression hosts, and DNA is normally introduced in a plasmid expression vector. The techniques for overexpression in *E. coli* are well developed and work by increasing the number of copies of the gene or increasing the binding strength of the promoter region so assisting transcription.

For example a DNA sequence for a protein of interest could be cloned or subcloned into a high copy-number plasmid containing the *lac* promoter, which is then transformed into the bacterium *Escherichia coli*. Addition of IPTG (a lactose analog) activates the lac promoter and causes the bacteria to express the protein of interest.

Corynebacterium

Non-pathogenic species of the gram-positive *Corynebacterium* are used for the commercial production of various amino acids. The *C. glutamicum* species is widely used for producing glutamate and lysine, components of human food, animal feed, and pharmaceutical products.

Expression of functionally active human epidermal growth factor has been done in *C. glutamicum*, thus demonstrating a potential for industrial-scale production of human proteins. Expressed proteins can be targeted for secretion through either the general secretory pathway (Sec) or the twin-arginine translocation pathway (Tat). Unlike gram-negative bacteria, the gram-positive *Corynebacterium* lack lipopolysaccharides that function as antigenic endotoxins in humans.

Yeast Systems

Expression systems in yeast typically use the common and well known *S.cerevisiae*, but also *Bacillus* gender. Systems using 'Pichia pastoris', that are licensed, allow stable and lasting production of proteins closer to mammalian cells, at high yield, in chemically defined media of proteins.

Insect Cell Systems

Insect cells (Sf9, Sf21, Hi-5 stains) allows expression of glycosylated proteins that cannot be expressed using *E.coli* or yeasts. It is very useful system for expression of proteins in high quantity. Genes are not expressed continuously because infected host cells will eventually lyse and die during each infection cycle. Non-lytic insect cell expression is an alternative to the lytic baculovirus expression system. In non-lytic expression, vectors are transiently or stably transfected into the chromosomal DNA of insect cells for subsequent protein expression. This is followed by selection and screening of recombinant clones. The non-lytic system has been used to give higher protein yield and quicker expression of recombinant proteins compared to baculovirus-infected cell expression. Cell lines used for this system include: Sf9, Sf21 from Spodoptera frugiperda cells, Hi-5 from Trichoplusia ni cells, and Schneider 2 cells and Schneider 3 cells from Drosophila melanogaster cells. With this system, cells do not lyse and several cultivation modes can be used. Additionally, protein production runs are reproducible. This system gives a homogeneous product. A drawback of this system is the requirement of an additional screening step of selecting viable clones.

Eukaryotic Systems

Protozoan *Leishmania tarentolae* (non pathogenic strain) expression system, also licensed, allow stable and lasting production of proteins at high yield, in chemically defined media. Produced proteins exhibit fully eukaryotic post-translational modifications, including glycosylation and disulfide bond formation.

Cell-Free Systems

Cell-free expression of proteins is performed in vitro using purified RNA polymerase, ribosomes, tRNA and ribonucleotides.

These reagents may be produced by extraction from cells or from a cell-based expression system. Due to the low expression levels and high cost of cell-free systems cell-based systems are more widely used.

Protein Biosynthesis

Protein biosynthesis is the process in which cells build or manufacture proteins. The term is sometimes used to refer only to protein translation but more often it refers to a multi-step process, beginning with amino acid synthesis and transcription of nuclear DNA into messenger RNA, which is then used as input for translation.

The cistron DNA is transcribed into a variety of RNA intermediates. The last version is used as a template in synthesis of a polypeptide chain. Proteins can often be synthesized directly from genes by translating mRNA.

When a protein must be available on short notice or in large quantities, a protein precursor is produced. A proprotein is an inactive protein containing one or more inhibitory peptides that can be activated when the inhibitory sequence is removed by proteolysis during post-translational modification.

A preprotein is a form that contains a signal sequence (an N-terminal signal peptide) that specifies its insertion into or through membranes, i.e., targets them for secretion. The signal peptide is cleaved off in the endoplasmic reticulum. Preproproteins have both sequences (inhibitory and signal) still present.

In protein synthesis, a succession of tRNA molecules charged with appropriate amino acids are brought together with an mRNA molecule and matched up by base-pairing through the anti-codons of the tRNA with successive codons of the mRNA. The amino acids are then linked together to extend the growing protein chain, and the tRNAs, no longer carrying amino acids, are released.

This whole complex of processes is carried out by the ribosome, formed of two main chains of RNA, called ribosomal RNA (rRNA), and more than 50 different proteins. The ribosome latches onto the end of an mRNA molecule and moves along it, capturing loaded tRNA molecules and joining together their amino acids to form a new protein chain. Protein biosynthesis, although very similar, is different for prokaryotes and eukaryotes.

Transcription

In transcription an mRNA chain is generated, with one strand of the DNA double helix in the genome as a template. This strand is called the template strand. Transcription can be divided into 3 stages: initiation, elongation, and termination, each regulated by a large number of proteins such as transcription factors and coactivators that ensure that the correct gene is transcribed.

The DNA strand is read in the 3' to 5' direction and the mRNA is transcribed in the 5' to 3' direction by the RNA polymerase. Transcription occurs in the cell nucleus, where the DNA is held. The DNA structure of the cell is made up of two helixes made up of sugar and phosphate held together by the bases. The sugar and the phosphate are joined together by a hydrogen bond. The DNA is "unzipped" by the enzyme helicase, leaving the single nucleotide chain open to be copied. RNA polymerase reads the DNA strand from 3-prime (3') end to the 5-prime (5') end, while it synthesizes a single strand of messenger RNA in the 5'-to-3' direction.

The general RNA structure is very similar to the DNA structure, but in RNA the nucleotide uracil takes the place that thymine occupies in DNA. The single strand of mRNA leaves the nucleus through nuclear pores, and migrates into the cytoplasm. The first product of transcription differs in prokaryotic cells from that of eukaryotic cells, as in prokaryotic cells the product is mRNA, which needs no post-transcriptional modification, whereas, in eukaryotic cells, the first product is called primary transcript, that needs post-transcriptional modification (capping with 7-methyl-guanosine, tailing with a poly A tail) to give hnRNA (heterophil nuclear RNA). hnRNA then undergoes splicing of introns (noncoding parts of the gene) via spliceosomes to produce the final mRNA.

Transcription is the process of creating a complementary RNA copy of a sequence of DNA. Both RNA and DNA are nucleic acids, which use base pairs of nucleotides as a complementary language that can be converted back and forth from DNA to RNA by the action of the correct enzymes. During transcription, a DNA sequence is read by an RNA polymerase, which produces a complementary, antiparallel RNA strand. As opposed to DNA replication, transcription results in an RNA complement that includes uracil (U) in all instances where thymine (T) would have occurred in a DNA complement. Also unlike DNA replication where DNA is synthesised, transcription does not involve an RNA primer to initiate RNA synthesis.

Transcription is explained easily in 4 or 5 steps, each moving like a wave along the DNA.

1. RNA polymerase moves the transcription bubble, a stretch of unpaired nucleotides, by breaking the hydrogen bonds between complementary nucleotides.
2. RNA polymerase adds matching RNA nucleotides that are paired with complementary DNA bases.
3. RNA sugar-phosphate backbone forms with assistance from RNA polymerase.
4. Hydrogen bonds of the untwisted RNA + DNA helix break, freeing the newly synthesized RNA strand.
5. If the cell has a nucleus, the RNA is further processed (addition of a 3' poly-A tail and a 5' cap) and exits through to the cytoplasm through the nuclear pore complex.

Transcription is the first step leading to gene expression. The stretch of DNA transcribed into an RNA molecule is called a *transcription unit* and encodes at least one gene.

If the gene transcribed encodes a protein, the result of transcription is messenger RNA (mRNA), which will then be used to create that protein via the process of translation. Alternatively, the transcribed gene may encode for either non-coding RNA genes (such as microRNA, lincRNA, etc.) or ribosomal RNA (rRNA) or transfer RNA (tRNA), other components of the protein-assembly process, or other ribozymes.

A DNA transcription unit encoding for a protein contains not only the sequence that will eventually be directly translated into the protein (the *coding sequence*) but also *regulatory sequences* that direct and regulate the synthesis of that protein. The regulatory sequence before (upstream from) the coding sequence is called the five prime untranslated region (5'UTR), and the sequence following (downstream from) the coding sequence is called the three prime untranslated region (3'UTR).

Transcription has some proofreading mechanisms, but they are fewer and less effective than the controls for copying DNA; therefore, transcription has a lower copying fidelity than DNA replication.

As in DNA replication, DNA is read from 3' $\rightarrow$ 5' during transcription. Meanwhile, the complementary RNA is created from the 5' $\rightarrow$ 3' direction. This means its 5' end is created first in base pairing. Although DNA is arranged as two antiparallel strands in a double helix, only one of the two DNA strands, called the template strand, is used for transcription.

This is because RNA is only single-stranded, as opposed to double-stranded DNA. The other DNA strand is called the coding (lagging) strand, because its sequence is the same as the newly created RNA transcript (except for the substitution of uracil for thymine). The use of only the 3' $\rightarrow$ 5' strand eliminates the need for the Okazaki fragments seen in DNA replication.

Transcription is divided into 5 stages: *pre-initiation, initiation, promoter clearance, elongation* and *termination.*

Pre-Initiation

In eukaryotes, RNA polymerase, and therefore the initiation of transcription, requires the presence of a core promoter sequence in the DNA. Promoters are regions of DNA that promote transcription and, in eukaryotes, are found at -30, -75, and -90 base pairs upstream from the transcription start site (abbreviated to TSS). Core promoters are sequences within the promoter that are essential for transcription initiation. RNA polymerase is able to bind to core promoters in the presence of various specific transcription factors.

The most characterised type of core promoter in eukaryotes is a short DNA sequence known as a TATA box, found 25-30 base pairs upstream from the TSS. The TATA box, as a core promoter, is the binding site for a transcription factor known as TATA-binding protein (TBP), which is itself a subunit of another transcription factor, called Transcription Factor II D (TFIID).

After TFIID binds to the TATA box via the TBP, five more transcription factors and RNA polymerase combine around the TATA box in a series of stages to form a preinitiation complex. One transcription factor, Transcription factor II H, has two components with helicase activity and so is involved in the separating of opposing strands of double-stranded DNA to form the initial transcription bubble.

However, only a low, or basal, rate of transcription is driven by the preinitiation complex alone. Other proteins known as activators and repressors, along with any associated coactivators or corepressors, are responsible for modulating transcription rate. Thus, preinitiation complex contains: 1. Core Promoter Sequence 2. Transcription Factors 3. RNA Polymerase 4. Activators and Repressors. The transcription preinitiation in archaea is, in essence, homologous to that of eukaryotes, but is much less complex. The archaeal preinitiation complex assembles at a TATA-box binding site; however, in archaea, this complex is composed of only RNA polymerase II, TBP, and TFB (the archaeal homologue of eukaryotic transcription factor II B (TFIIB)).

Initiation

In bacteria, transcription begins with the binding of RNA polymerase to the promoter in DNA. RNA polymerase is a core enzyme consisting of five subunits: 2 α subunits, 1 β subunit, 1 β' subunit, and 1 ω subunit. At the start of initiation, the core enzyme is associated with a sigma factor that aids in finding the appropriate -35 and -10 base pairs downstream of promoter sequences. When the sigma factor and RNA polymerase combine, they form a holoenzyme.

Transcription initiation is more complex in eukaryotes. Eukaryotic RNA polymerase does not directly recognise the core promoter sequences. Instead, a collection of proteins called transcription factors mediate the binding of RNA polymerase and the initiation of transcription. Only after certain transcription factors are attached to the promoter does the RNA polymerase bind to it. The completed assembly of transcription factors and RNA polymerase bind to the promoter, forming a transcription initiation complex. Transcription in the archaea domain is similar to transcription in eukaryotes.

Promoter Clearance

After the first bond is synthesized, the RNA polymerase must clear the promoter. During this time there is a tendency to release the RNA transcript and produce truncated transcripts.

This is called *abortive initiation* and is common for both eukaryotes and prokaryotes. Abortive initiation continues to occur until the σ factor rearranges, resulting in the transcription elongation complex (which gives a 35 bp moving footprint). The σ factor is released before 80 nucleotides of mRNA are synthesized.

Once the transcript reaches approximately 23 nucleotides, it no longer slips and elongation can occur. This, like most of the remainder of transcription, is an energy-dependent process, consuming adenosine triphosphate (ATP).

Promoter clearance coincides with phosphorylation of serine 5 on the carboxy terminal domain of RNAP II in eukaryotes, which is phosphorylated by TFIIH.

Elongation

One strand of the DNA, the *template strand* (or noncoding strand), is used as a template for RNA synthesis. As transcription proceeds, RNA polymerase traverses the template strand and uses base pairing complementarity with the DNA template to create an RNA copy.

Although RNA polymerase traverses the template strand from 3' → 5', the coding (non-template) strand and newly-formed RNA can also be used as reference points, so transcription can be described as occurring 5' → 3'.

This produces an RNA molecule from 5' → 3', an exact copy of the coding strand (except that thymines are replaced with uracils, and the nucleotides are composed of a ribose (5-carbon) sugar where DNA has deoxyribose (one less oxygen atom) in its sugar-phosphate backbone).

Unlike DNA replication, mRNA transcription can involve multiple RNA polymerases on a single DNA template and multiple rounds of transcription (amplification of particular mRNA), so many mRNA molecules can be rapidly produced from a single copy of a gene. Elongation also involves a proofreading mechanism that can replace incorrectly incorporated bases. In eukaryotes, this may correspond with short pauses during transcription that allow appropriate RNA editing factors to bind. These pauses may be intrinsic to the RNA polymerase or due to chromatin structure.

Termination

Bacteria use two different strategies for transcription termination.1.Rho-independent transcription 2.Rho-dependent transcription. In Rho-independent transcription termination, also called intrinsic termination, RNA transcription stops when the newly synthesized RNA molecule forms a G-C-rich hairpin loop followed by a run of Us. When the hairpin forms, the mechanical stress breaks the weak rU-dA bonds, now filling the DNA-RNA hybrid. This pulls the poly-U transcript out of the active site of the RNA polymerase, in effect, terminating transcription. In the "Rho-dependent" type of termination, a protein factor called "Rho" destabilizes the interaction between the template and the mRNA, thus releasing the newly synthesized mRNA from the elongation complex.

Transcription termination in eukaryotes is less understood but involves cleavage of the new transcript followed by template-independent addition of *A*s at its new 3' end, in a process called polyadenylation.

Measuring and Detecting Transcription

Transcription can be measured and detected in a variety of ways:

- Nuclear Run-on assay: measures the relative abundance of newly formed transcripts
- RNase protection assay and ChIP-Chip of RNAP: detect active transcription sites
- RT-PCR: measures the absolute abundance of total or nuclear RNA levels, which may however differ from transcription rates
- DNA microarrays: measures the relative abundance of the global total or nuclear RNA levels; however, these may differ from transcription rates
- In situ hybridization: detects the presence of a transcript
- MS2 tagging: by incorporating RNA stem loops, such as MS2, into a gene, these become incorporated into

newly synthesized RNA. The stem loops can then be detected using a fusion of GFP and the MS2 coat protein, which has a high affinity, sequence-specific interaction with the MS2 stem loops. The recruitment of GFP to the site of transcription is visualised as a single fluorescent spot. This remarkable new approach has revealed that transcription occurs in discontinuous bursts, or pulses. With the notable exception of in situ techniques, most other methods provide cell population averages, and are not capable of detecting this fundamental property of genes.

- Northern blot: the traditional method, and until the advent of RNA-Seq, the most quantitative
- RNA-Seq: applies next-generation sequencing techniques to sequence whole transcriptomes, which allows the measurement of relative abundance of RNA, as well as the detection of additional variations such as fusion genes, post-translational edits and novel splice sites

Transcription Factories

Active transcription units are clustered in the nucleus, in discrete sites called transcription factories or euchromatin. Such sites can be visualized by allowing engaged polymerases to extend their transcripts in tagged precursors (Br-UTP or Br-U) and immuno-labelling the tagged nascent RNA. Transcription factories can also be localized using fluorescence in situ hybridization or marked by antibodies directed against polymerases. There are ~10,000 factories in the nucleoplasm of a HeLa cell, among which are ~8,000 polymerase II factories and ~2,000 polymerase III factories.

Each polymerase II factory contains ~8 polymerases. As most active transcription units are associated with only one polymerase, each factory usually contains ~8 different transcription units. These units might be associated through promoters and/or enhancers, with loops forming a 'cloud' around the factor. In genetics, a transcription factory is an active gene transcription unit that is clustered in a discrete site within the eukaryotic nucleus. Such sites can be visualized by allowing engaged polymerases to extend their transcripts with tagged precursors (Br-UTP or Br-U) and immuno-labelling the tagged nascent RNA. Transcription factories can also be detected using fluorescence *in situ* hybridization or marked by antibodies directed against RNA polymerases. There are ~10,000 factories in the nucleoplasm of a HeLa cell, which reflect ~8,000 RNA polymerase II holoenzyme factories and ~2,000 RNA polymerase III factories. As most active transcription units are associated with only one polymerase, one estimate suggests a polymerase II factory contains on average ~8 polymerase holoenzymes. The nucleolus is often considered to be a large RNA polymerase I factory, although this structure is the site of other activities than just rRNA transcription.

RNA Polymerase II Holoenzyme Transcription Factory

There are ~8,000 RNA polymerase II holoenzyme factories in the nucleoplasm of a HeLa cell, but only 100-300 RNAP II foci per nucleus in erythroid cells like many other tissue types. The number of transcription factories in tissues is far more restricted than indicated by previous estimates from cultured cells. As an active transcription unit is usually associated with only one Pol II holoenzyme, a polymerase II factory may contain on average ~8 holoenzymes. Colocalization of transcribed genes has not been observed when using cultured fibroblast-like cells. Differentiated or committed tissue types have a limited number of available transcription sites. Estimates show that erythroid cells express at least 4,000 genes, so many genes are obliged to seek out and share the same factory.

Gene Intranuclear Position

The intranuclear position of many genes is correlated with their activity state. During transcription *in vivo*, distal active genes are dynamically organised into shared nuclear subcompartments and colocalize to the same transcription factory at high frequencies. Movement into or out of these factories results in activation (On) or abatement (Off) of transcription, rather than by recruiting and assembling a transcription complex. Usually, genes migrate to preassembled factories for transcription. An expressed gene is preferentially located outside of its chromosome territory, but a closely linked, inactive gene is located inside. A molecule that allows the genetic material to be realised as a protein was first hypothesized by Fraŋŋois Jacob and Jacques Monod. Severo Ochoa won a Nobel Prize in Physiology or Medicine for developing a process of RNA synthesis in 1959.

RNA synthesis by RNA polymerase was established *in vitro* by several laboratories by 1965; however, the RNA synthesized by these enzymes had properties that suggested the existence of an additional factor needed to terminate transcription correctly.

In 1972, Walter Fiers became the first person to actually prove the existence of the terminating enzyme. Roger D. Kornberg won the 2006 Nobel Prize in Chemistry "for his studies of the molecular basis of eukaryotic transcription".

Reverse Transcription

Some viruses (such as HIV, the cause of AIDS), have the ability to transcribe RNA into DNA. HIV has an RNA genome that is duplicated into DNA. The resulting DNA can be merged with the DNA genome of the host cell. The main enzyme responsible for synthesis of DNA from an RNA template is called reverse transcriptase. In the case of HIV, reverse transcriptase is responsible for synthesizing a complementary DNA strand (cDNA) to the viral RNA genome. An associated enzyme, ribonuclease H, digests the RNA strand, and reverse transcriptase synthesises a complementary strand of DNA to form a double helix DNA structure. This cDNA is integrated into the host cell's genome via another enzyme (integrase) causing the host cell to generate viral proteins that reassemble into new viral particles. In HIV, subsequent to this, the host cell undergoes programmed cell death, apoptosis of T cells. However, in other retroviruses, the host cell remains intact as the virus buds out of the cell. Some eukaryotic cells contain an enzyme with reverse transcription activity called telomerase. Telomerase is a reverse transcriptase that lengthens the ends of linear chromosomes. Telomerase carries an RNA template from which it synthesizes DNA repeating sequence, or "junk" DNA.

This repeated sequence of DNA is important because, every time a linear chromosome is duplicated, it is shortened in length. With "junk" DNA at the ends of chromosomes, the shortening eliminates some of the non-essential, repeated sequence rather than the protein-encoding DNA sequence farther away from the chromosome end.

Telomerase is often activated in cancer cells to enable cancer cells to duplicate their genomes indefinitely without losing important protein-coding DNA sequence. Activation of telomerase could be part of the process that allows cancer cells to become *immortal*. However, the true *in vivo* significance of telomerase has still not been empirically proven.

Inhibitors

Transcription inhibitors can be used as antibiotics against, for example, pathogenic bacteria (antibacterials) and fungi (antifungals). An example of such an antibacterial is rifampicin, which inhibits prokaryotic DNA transcription into mRNA by inhibiting DNA-dependent RNA polymerase by binding its beta-subunit. 8-Hydroxyquinoline is an antifungal transcription inhibitor.

Translation

The synthesis of proteins is known as translation. Translation occurs in the cytoplasm, where the ribosomes are located. Ribosomes are made of a small and large subunit that surround the mRNA. In translation, messenger RNA (mRNA) is decoded to produce a specific polypeptide according to the rules specified by the trinucleotide genetic code.

This uses an mRNA sequence as a template to guide the synthesis of a chain of amino acids that form a protein. Translation proceeds in four phases: activation, initiation, elongation, and termination (all describing the growth of the amino acid chain, or polypeptide that is the product of translation).

In activation, the correct amino acid (AA) is joined to the correct transfer RNA (tRNA). While this is not, in the technical sense, a step in translation, it is required for translation to proceed. The AA is joined by its carboxyl group to the 3' OH of the tRNA by an ester bond. When the tRNA has an amino acid linked to it, it is termed "charged". Initiation involves the small subunit of the ribosome binding to 5' end of mRNA with the help of initiation factors (IF), other proteins that assist the process. Elongation occurs when the next aminoacyl-tRNA (charged tRNA) in line binds to the ribosome along with GTP and an elongation factor.

Termination of the polypeptide happens when the A site of the ribosome faces a stop codon (UAA, UAG, or UGA). When this happens, no tRNA can recognise it, but releasing factor can recognise nonsense codons and causes the release of the polypeptide chain.

The capacity of disabling or inhibiting translation in protein biosynthesis is used by some antibiotics such as anisomycin, cycloheximide, chloramphenicol, tetracycline, streptomycin, erythromycin, puromycin, etc.

In molecular biology and genetics, translation is the third stage of protein biosynthesis (part of the overall process of gene expression). In translation, messenger RNA (mRNA) produced by transcription is decoded by the ribosome to produce a specific amino acid chain, or polypeptide, that will later fold into an active protein.

In Bacteria, translation occurs in the cell's cytoplasm, where the large and small subunits of the ribosome are located, and bind to the mRNA. In Eukaryotes, translation occurs across the membrane of the endoplasmic reticulum in a process called vectorial synthesis. The ribosome facilitates decoding by inducing the binding of tRNAs with complementary anticodon sequences to that of the mRNA. The tRNAs carry specific amino acids that are chained together into a polypeptide as the mRNA passes through and is "read" by the ribosome in a fashion reminiscent to that of a stock ticker and ticker tape.

In many instances, the entire ribosome/mRNA complex causing it to bind to the outer membrane of the rough endoplasmic reticulum and release the nascent protein polypeptide inside for later vesicle transport and secretion outside of the cell. Many types of transcribed RNA, such as transfer RNA, ribosomal RNA, and small nuclear RNA, do not undergo translation into proteins.

Translation proceeds in four phases: initiation, elongation, translocation and termination (all describing the growth of the amino acid chain, or polypeptide that is the product of translation). Amino acids are brought to ribosomes and assembled into proteins.

In activation, the correct amino acid is covalently bonded to the correct transfer RNA (tRNA). The amino acid is joined by its carboxyl group to the 3' OH of the tRNA by an ester bond. When the tRNA has an amino acid linked to it, it is termed "charged". Initiation involves the small subunit of the ribosome binding to the 5' end of mRNA with the help of initiation factors (IF). Termination of the polypeptide happens when the A site of the ribosome faces a stop codon (UAA, UAG, or UGA). No tRNA can recognise or bind to this codon. Instead, the stop codon induces the binding of a release factor protein that prompts the disassembly of the entire ribosome/mRNA complex. A number of antibiotics act by inhibiting translation; these include anisomycin, cycloheximide, chloramphenicol, tetracycline, streptomycin, erythromycin, and puromycin, among others. Prokaryotic ribosomes have a different structure from that of eukaryotic ribosomes, and thus antibiotics can specifically target bacterial infections without any detriment to a eukaryotic host's cells.

The basic process of protein production is addition of one amino acid at a time to the end of a protein. This operation is performed by a ribosome. The choice of amino acid type to add is determined by an mRNA molecule. Each amino acid added is matched to a three nucleotide subsequence of the mRNA.

For each such triplet possible, only one particular amino acid type is accepted. The successive amino acids added to the chain are matched to successive nucleotide triplets in the mRNA. In this way the sequence of nucleotides in the template mRNA chain determines the sequence of amino acids in the generated amino acid chain. Addition of an amino acid occurs at the C-terminus of the peptide and thus translation is said to be amino-to-carboxyl directed. The mRNA carries genetic information encoded as a ribonucleotide sequence from the chromosomes to the ribosomes. The ribonucleotides are "read" by translational machinery in a sequence of nucleotide triplets called codons. Each of those triplets codes for a specific amino acid. The ribosome molecules translate this code to a specific sequence of amino acids. The ribosome is a multisubunit structure containing rRNA and proteins. It is the "factory" where amino acids are assembled into proteins. tRNAs are small noncoding RNA chains (74-93 nucleotides) that transport amino acids to the ribosome. tRNAs have a site for amino acid attachment, and a site called an anticodon. The anticodon is an RNA triplet complementary to the mRNA triplet that codes for their cargo amino acid.

Aminoacyl tRNA synthetase (an enzyme) catalyses the bonding between specific tRNAs and the amino acids that their anticodon sequences call for. The product of this reaction is an aminoacyl-tRNA molecule.

This aminoacyl-tRNA travels inside the ribosome, where mRNA codons are matched through complementary base pairing to specific tRNA anticodons. The ribosome has three sites for tRNA to bind. They are the aminoacyl site (abbreviated A), the peptidyl site (abbreviated P) and the exit site (abbreviated E). With respect to the mRNA, the three sites are oriented 5'to 3' E-P-A, because ribosomes moves toward the 3' end of mRNA. The A site binds the incoming tRNA with the complementary codon on the mRNA. The P site holds the tRNA with the growing polypeptide chain. The E site holds the tRNA without its amino acid. When an aminoacyl-tRNA initially binds to its corresponding codon on the mRNA, it is in the A site. Then, a peptide bond forms between the amino acid of the tRNA in the A site and the amino acid of the charged tRNA in the P site. The growing polypeptide chain is transferred to the tRNA in the A site.

Translocation occurs, moving the tRNA in the P site, now without an amino acid, to the E site; the tRNA that was in the A site, now charged with the polypeptide chain, is moved to the P site. The tRNA in the E site leaves and another aminoacyl-tRNA enters the A site to repeat the process. After the new amino acid is added to the chain, the energy provided by the hydrolysis of a GTP bound to the translocase EF-G (in prokaryotes) and eEF-2 (in eukaryotes) moves the ribosome down one codon towards the 3' end. The energy required for translation of proteins is significant. For a protein containing n amino acids, the number of high-energy Phosphate bonds required to translate it is $4n$-1. The rate of translation varies; it is significantly higher in prokaryotic cells (up to 17-21 amino acid residues per second) than in eukaryotic cells (up to 6-9 amino acid residues per second).

Genetic Code

Whereas other aspects such as the 3D structure, called tertiary structure, of protein can only be predicted using sophisticated algorithms, the amino acid sequence, called primary structure, can be determined solely from the nucleic acid sequence with the aid of a translation table.

This approach may not give the correct amino acid composition of the protein, in particular if unconventional amino acids such as selenocysteine are incorporated into the protein, which is coded for by a conventional stop codon in combination with a downstream hairpin (SElenoCysteine Insertion Sequence, or SECIS).

There are many computer programs capable of translating a DNA/RNA sequence into a protein sequence. Normally this is performed using the Standard Genetic Code; many bioinformaticians have written at least one such program at some point in their education. However, few programs can handle all the "special" cases, such as the use of the alternative initiation codons. For instance, the rare alternative start codon CTG codes for Methionine when used as a start codon, and for Leucine in all other positions.

Example: Condensed translation table for the Standard Genetic Code (from the NCBI Taxonomy webpage).

Translation Tables

Even when working with ordinary Eukaryotic sequences such as the Yeast genome, it is often desired to be able to use alternative translation tables—namely for translation of the mitochondrial genes.

Events Following Protein Translation

The events following biosynthesis include post-translational modification and protein folding. During and after synthesis, polypeptide chains often fold to assume, so called, native secondary and tertiary structures. This is known as *protein folding*. Post-translational modification (PTM) is the chemical modification of a protein after its translation. It is one of the later steps in protein biosynthesis, and thus gene expression, for many proteins. A protein (also called a polypeptide) is a chain of amino acids. During protein synthesis, 20 different amino acids can be incorporated to become a protein. After translation, the Post-translational modification of amino acids extends the range of functions of the protein by attaching it to other biochemical functional groups (such as acetate, phosphate, various lipids and carbohydrates), changing the chemical nature of an amino acid (e.g. citrullination), or making structural changes (e.g. formation of disulfide bridges).

Also, enzymes may remove amino acids from the amino end of the protein, or cut the peptide chain in the middle. For instance, the peptide hormone insulin is cut twice after disulfide bonds are formed, and a propeptide is removed from the middle of the chain; the resulting protein consists of two polypeptide chains connected by disulfide bonds. Also, most nascent polypeptides start with the amino acid methionine because the "start" codon on mRNA also codes for this amino acid. This amino acid is usually taken off during post-translational modification. Other modifications, like phosphorylation, are part of common mechanisms for controlling the behaviour of a protein, for instance activating or inactivating an enzyme.

Post-translational modification of proteins is detected by mass spectrometry or Eastern blotting.

Post-Translational Modification Statistics

Recently statistics of each post-translational modification experimentally and putatively detected have been compiled using proteome-wide information from the Swiss-Prot database.

Protein folding is the process by which a protein structure assumes its functional shape or conformation. It is the physical process by which a polypeptide folds into its characteristic and functional three-dimensional structure from random coil. Each protein exists as an unfolded polypeptide or random coil when translated from a sequence of mRNA to a linear chain of amino acids. This polypeptide lacks any developed three-dimensional structure. Amino acids interact with each other to produce a well-defined three-dimensional structure, the folded protein, known as the native state. The resulting three-dimensional structure is determined by the amino acid sequence (Anfinsen's dogma). The correct three-dimensional structure is essential to function, although some parts of functional proteins may remain unfolded. Failure to fold into native structure produces inactive proteins that are usually toxic. Several neurodegenerative and other diseases are believed to result from the accumulation of amyloid fibrills formed by *misfolded* proteins. Many allergies are caused by the folding of the proteins, for the immune system does not produce antibodies for certain protein structures.

Relationship between Folding and Amino Acid Sequence

The amino-acid sequence of a protein determines its native conformation. A protein molecule folds spontaneously during or after biosynthesis. While these macromolecules may be regarded as "folding themselves", the process also depends on the solvent (water or lipid bilayer), the concentration of salts, the temperature, and the presence of molecular chaperones. Folded proteins usually have a side chain. Packing stabilizes the folded state, and charged or side chains occupy the solvent-exposed surface where they interact with surrounding water. Minimising the number of hydrophobic side-chains exposed to water is an important driving force behind the folding process.

Formation of intramolecular hydrogen bonds provides another important contribution to protein stability. The strength of hydrogen bonds depends on their environment, thus H-bonds enveloped in a hydrophobic core contribute more than H-bonds exposed to the aqueous environment to the stability of the native state.

The process of folding often begins co-translationally, so that the N-terminus of the protein begins to fold while the C-terminal portion of the protein is still being synthesized by the ribosome. Specialised proteins called chaperones assist in the folding of other proteins. A well studied example is the bacterial GroEL system, which assists in the folding of globular proteins.

In eukaryotic organisms chaperones are known as heat shock proteins. Although most globular proteins are able to assume their native state unassisted, chaperone-assisted folding is often necessary in the crowded intracellular environment to prevent aggregation; chaperones are also used to prevent misfolding and aggregation that may occur as a consequence of exposure to heat or other changes in the cellular environment.

There are two models of protein folding that are currently being confirmed: The first: The diffusion collision model, in which a nucleus is formed, then the secondary structure is formed, and finally these secondary structures are collided together and pack tightly together.

The second: The nucleation-condensation model, in which the secondary and tertiary structures of the protein are made at the same time. Recent studies have shown that some proteins show characteristics of both of these folding models.

For the most part, scientists have been able to study many identical molecules folding together *en masse*. At the coarsest level, it appears that in transitioning to the native state, a given amino acid sequence takes on roughly the same route and proceeds through roughly the same intermediates and transition states.

Often folding involves first the establishment of regular secondary and supersecondary structures, in particular alpha helices and beta sheets, and afterward tertiary structure. Formation of quaternary structure usually involves the "assembly" or "coassembly" of subunits that have already folded. The regular alpha helix and beta sheet structures fold rapidly because they are stabilized by intramolecular hydrogen bonds, as was first characterised by Linus Pauling. Protein folding may involve covalent bonding in the form of disulfide bridges formed between two cysteine residues or the formation of metal clusters. Shortly before settling into their more energetically favourable native conformation, molecules may pass through an intermediate "molten globule" state.

The essential fact of folding, however, remains that the amino acid sequence of each protein contains the information that specifies both the native structure and the pathway to attain that state.

This is not to say that nearly identical amino acid sequences always fold similarly. Conformations differ based on environmental factors as well; similar proteins fold differently based on where they are found. Folding is a spontaneous process independent of energy inputs from nucleoside triphosphates.

The passage of the folded state is mainly guided by hydrophobic interactions, formation of intramolecular hydrogen bonds, and van der Waals forces, and it is opposed by conformational entropy.

Disruption of the Native State

Under some conditions proteins will not fold into their biochemically functional forms. Temperatures above or below the range that cells tend to live in will cause thermally unstable proteins to unfold or "denature" (this is why boiling makes an egg white turn opaque).

High concentrations of solutes, extremes of pH, mechanical forces, and the presence of chemical denaturants can do the same. Protein thermal stability is far from constant, however. For example, hyperthermophilic bacteria have been found that grow at temperatures as high as 122 °C, which of course requires that their full complement of vital proteins and protein assemblies be stable at that temperature or above.

A fully denatured protein lacks both tertiary and secondary structure, and exists as a so-called random coil. Under certain conditions some proteins can refold; however, in many cases, denaturation is irreversible.

Cells sometimes protect their proteins against the denaturing influence of heat with enzymes known as chaperones or heat shock proteins, which assist other proteins both in folding and in remaining folded. Some proteins never fold in cells at all except with the assistance of chaperone molecules, which either isolate individual proteins so that their folding is not interrupted by interactions with other proteins or help to unfold misfolded proteins, giving them a second chance to refold properly. This function is crucial to prevent the risk of precipitation into insoluble amorphous aggregates.

Incorrect Protein Folding and Neurodegenerative Disease

Aggregated proteins are associated with prion-related illnesses such as Creutzfeldt-Jakob disease, bovine spongiform encephalopathy (mad cow disease), amyloid-related illnesses such as Alzheimer's disease and familial amyloid cardiomyopathy or polyneuropathy, as well as intracytoplasmic aggregation diseases such as Huntington's and Parkinson's disease.

These age onset degenerative diseases are associated with the aggregation of misfolded proteins into insoluble, extracellular aggregates and/or intracellular inclusions including cross-beta sheet amyloid fibrils.

While it is not completely clear whether the aggregates are the cause or merely a reflection of the loss of protein homeostasis, the balance between synthesis, folding, aggregation and protein turnover, the recent European Medicines Agency approval of Tafamidis or Vyndaqel (a kinetic stabilizer of tetrameric transthyretin) for the treatment of the transthyretin amyloid diseases suggests that it is the process of amyloid fibril formation and not the fibrils themselves that causes the degeneration of post-mitotic tissue in human amyloid diseases.

Misfolding and excessive degradation instead of folding and function leads to a number of proteopathy diseases such as antitrypsin-associated emphysema, cystic fibrosis and the lysosomal storage diseases, where loss of function is the origin of the disorder. While protein replacement therapy has historically been used to correct the latter disorders, an emerging approach is to use pharmaceutical chaperones to fold mutated proteins to render them functional.

Effect of External Factors on the Folding of Proteins

Several external factors such as temperature, external fields (electric, magnetic), molecular crowding, limitation of space could have a big influence on the folding of proteins. Modification of the local minima by external factors can also induce modifications of the folding trajectory.

Protein folding is a very finely tuned process. Hydrogen bonding between different atoms provides the force required. Hydrophobic interactions between hydrophobic amino acids pack the hydrophobic residues.

The Levinthal Paradox and Kinetics

Levinthal's paradox is a thought experiment, also constituting a self-reference in the theory of protein folding. In 1969, Cyrus Levinthal noted that, because of the very large number of degrees of freedom in an unfolded polypeptide chain, the molecule has an astronomical number of possible conformations. An estimate of 3^{300} or 10^{143} was made in one of his papers. The Levinthal paradox observes that if a protein were folded by sequentially sampling of all possible conformations, it would take an astronomical amount of time to do so, even if the conformations were sampled at a rapid rate (on the nanosecond or picosecond scale). Based upon the observation that proteins fold much faster than this, Levinthal then proposed that a random conformational search does not occur, and the protein must, therefore, fold through a series of meta-stable intermediate states.

The duration of the folding process varies dramatically depending on the protein of interest. When studied outside the cell, the slowest folding proteins require many minutes or hours to fold primarily due to proline isomerization, and must pass through a number of intermediate states, like checkpoints, before the process is complete.

On the other hand, very small single-domain proteins with lengths of up to a hundred amino acids typically fold in a single step. Time scales of milliseconds are the norm and the very fastest known protein folding reactions are complete within a few microseconds.

Experimental Techniques for Studying Protein Folding

While inferences about protein folding can be made through mutation studies; typically, experimental techniques for studying protein folding rely on the gradual unfolding or folding of a solution of proteins and observing conformational changes using standard non-crystallographic techniques for observing protein structure.

Chapter 7

Natural Product Chemistry

A natural product is a chemical compound or substance produced by a living organism - found in nature that usually has a pharmacological or biological activity for use in pharmaceutical drug discovery and drug design. A *natural* product can be considered as such even if it can be prepared by total synthesis. These small molecules provide the source or inspiration for the majority of FDA-approved agents and continue to be one of the major sources of inspiration for drug discovery. In particular, these compounds are important in the treatment of life-threatening conditions.

Natural products may be extracted from tissues of terrestrial plants, marine organisms or microorganism fermentation broths. A crude (untreated) extract from any one of these sources typically contains novel, structurally diverse chemical compounds, which the natural environment is a rich source of. Chemical diversity in nature is based on biological and geographical diversity, so researchers travel around the world obtaining samples to analyse and evaluate in drug discovery screens or bioassays. This effort to search for natural products is known as bioprospecting.

Screening of Natural Products

Pharmacognosy provides the tools to identify, select and process natural products destined for medicinal use. Usually, the natural product compound has some form of biological activity and that compound is known as the active principle - such a structure can act as a lead compound (not to be confused with compounds containing the element lead). Many of today's medicines are obtained *directly* from a natural source. On the other hand, some medicines are *developed from a lead compound originally* obtained from a natural source. This means the lead compound:

- can be produced by total synthesis, or
- can be a starting point (precursor) for a semisynthetic compound, or
- can act as a template for a structurally different total synthetic compound.

This is because most biologically active natural product compounds are secondary metabolites with very complex structures.

This has an advantage in that they are extremely novel compounds but this complexity also makes many lead compounds' synthesis difficult and the compound usually has to be extracted from its natural source - a slow, expensive and inefficient process. As a result, there is usually an advantage in designing simpler analogues.

The Plant Kingdom

Plants have always been a rich source of lead compounds (e.g. Alkaloids, morphine, cocaine, digitalis, quinine, tubocurarine, nicotine, and muscarine). Many of these lead compounds are useful drugs in themselves (e.g. Alkaloids, morphine and quinine), and others have been the basis for synthetic drugs (e.g. local anaesthetics developed from cocaine).

Clinically useful drugs which have been recently isolated from plants include the anticancer agent paclitaxel (Taxol) from the yew tree, and the antimalarial agent artemisinin from *Artemisia annua*. Plants provide a large bank of rich, complex and highly varied structures which are unlikely to be synthesized in laboratories. Furthermore, evolution has already carried out a screening process itself whereby plants are more likely to survive if they contain potent compounds which deter animals from eating them. Even today, the number of plants that have been extensively studied is relatively very few and the vast majority have not been studied at all. Major classes of molecules include terpenoids, phytosterols, alkaloids, natural phenols and polyphenols.

The Microbial World

Microorganisms such as bacteria and fungi have been invaluable for discovering drugs and lead compounds. These microorganisms produce a large variety of antimicrobial agents which have evolved to give their hosts an advantage over their competitors in the microbiological world.

The screening of microorganisms became highly popular after the discovery of penicillin. Soil and water samples were collected from all over the world in order to study new bacterial or fungal strains, leading to an impressive arsenal of antibacterial agents such as the cephalosporins, tetracyclines, aminoglycosides, rifamycins, and chloramphenicol. Although most of the drugs derived from microorganisms are used in antibacterial therapy, some microbial metabolites have provided lead compounds in other fields of medicine.

For example, asperlicin - isolated from *Aspergillus alliaceus* - is a novel antagonist of a peptide hormone called cholecystokinin (CCK) which is involved in the control of appetite.

CCK also acts as a neurotransmitter in the brain and is thought to be involved in panic attacks. Analogues of asperlicin may therefore have potential in treating anxiety.

Other examples include the fungal metabolite lovastatin, which was the lead compound for a series of drugs that lower cholesterol levels, and another fungal metabolite called ciclosporin which is used to suppress the immune response after transplantation operations.

The Marine World

In recent years, there has been a great interest in finding lead compounds from marine sources. Coral, sponges, fish, and marine microorganisms have a wealth of biologically potent chemicals with interesting inflammatory, antiviral, and anticancer activity.

For example, curacin A is obtained from a marine cyanobacterium and shows potent antitumor activity. Other antitumor agents derived from marine sources include eleutherobin, discodermolide, bryostatins, dolostatins, and cephalostatins.

Animal Sources

Animals can sometimes be a source of new lead compounds. For example, a series of antibiotic peptides were extracted from the skin of the African clawed frog and a potent analgesic compound called epibatidine was obtained from the skin extracts of the Ecuadorian poison frog.

Venoms and Toxins

Venoms and toxins from animals, plants, snakes, spiders, scorpions, insects, and microorganisms are extremely potent because they often have very specific interactions with a macromolecular target in the body. As a result, they have proved important tools in studying receptors, ion channels, and enzymes. Many of these toxins are polypeptides (e.g. α-bungarotoxin from cobras).

However, non-peptide toxins such as tetrodotoxin from the puffer fish are also extremely potent. Venoms and toxins have been used as lead compounds in the development of novel drugs. For example, teprotide, a peptide isolated from the venom of the Brazilian viper, was the lead compound for the development of the antihypertensive agents cilazapril and captopril. The neurotoxins from *Clostridium botulinum* are responsible for serious food poisoning (botulism), but they have a clinical use as well.

They can be injected into specific muscles (such as those controlling the eyelid) to prevent muscle spasm. These toxins prevent cholinergic transmission and could well prove a lead for the development of novel anticholinergic drugs.

Traditional Medicine

In the past, traditional peoples or ancient civilizations depended greatly on local flora and fauna for their survival. They would experiment with various berries, leaves, roots, animal parts or minerals to find out what effects they had. As a result, many crude drugs were observed by the local healer or shaman to have some medical use. Although some preparations may have been dangerous, or worked by a ceremonial or placebo effect, traditional healing systems usually had a substantial active pharmacopoeia, and in fact most western medicines up until the 1920s were developed this way. Some systems, like traditional Chinese medicine or Ayurveda were fully as sophisticated and as documented systems as western medicine, although they might use different paradigms.

Many of these aqueous, ethanolic, distilled, condensed or dried extracts do indeed have a real and beneficial effect, and a study of ethnobotany can give clues as to which plants might be worth studying in more detail. Rhubarb root has been used as a purgative for many centuries.

In China, it was called "The General" because of its "galloping charge" and was only used for one or two doses unless processed to reduce its purgative qualities. (Bulk laxatives would follow or be used on weaker patients according to the complex laxative protocols of the medical system.) The most significant chemicals in rhubarb root are anthraquinones, which were used as the lead compounds in the design of the laxative dantron.

Insects have also gained recent attention as valuable sources of natural products and their use in traditional medicine has been reviewed.

The extensive records of Chinese medicine about response to Artemisia preparations for malaria also provided the clue to the novel antimalarial drug artemisinin.

The therapeutic properties of the opium poppy (active principle morphine) were known in Ancient Egypt, were those of the *Solanaceae* plants in ancient Greece (active principles atropine and hyoscine). The snakeroot plant was well regarded in India (active principle reserpine), and herbalists in medieval England used extracts from the willow tree(salicin) and foxglove (active principle digitalis - a mixture of compounds such as digitoxin, digitonin, digitalin). The Aztec and Mayan cultures of Mesoamerica used extracts from a variety of bushes and trees including the ipecacuanha root (active principle emetine), coca bush (active principle cocaine), and cinchona bark (active principle quinine).

It can be challenging to obtain information from practitioners of traditional medicine unless a genuine long term relationship is made. Ethnobotanist Richard Schultes approached the Amazonian shamans with respect, dealing with them on their terms. He became a "depswa" - medicine man

- sharing their rituals while gaining knowledge. They responded to his inquiries in kind, leading to new medicines. On the other hand Cherokee herbalist David Winston recounts how his uncle, a medicine priest, would habitually give misinformation to the visiting ethno-botanists.

The acupuncturists who investigated Mayan medicine recounted in Wind in the Blood had something to share with the native healers and thus were able to find information not available to anthropologists. The issue of rights to medicine derived from native plants used and frequently cultivated by native healers complicates this issue.

Isolation and Purification

If the lead compound (or active principle) is present in a mixture of other compounds from a natural source, it has to be isolated and purified. The ease with which the active principle can be isolated and purified depends much on the structure, stability, and quantity of the compound. For example, Alexander Fleming recognised the antibiotic qualities of penicillin and its remarkable non-toxic nature to humans, but he disregarded it as a clinically useful drug because he was unable to *purify* it. He could *isolate* it in aqueous solution, but whenever he tried to remove the water, the drug was destroyed.

It was not until the development of new experimental procedures such as freeze drying and chromatography that the successful isolation and purification of penicillin and other natural products became feasible.

Synthesis

Not all natural products can be fully synthesized and many natural products have very complex structures that are too difficult and expensive to synthesize on an industrial scale.

These include drugs such as penicillin, morphine, and paclitaxel (Taxol). Such compounds can only be harvested from their natural source - a process which can be tedious, time consuming, and expensive, as well as being wasteful on the natural resource. For example, one yew tree would have to be cut down to extract enough paclitaxel from its bark for a single dose. Furthermore, the number of structural analogues that can be obtained from harvesting is severely limited.

A further problem is that isolates often work differently than the original natural products which have synergies and may combine, say, antimicrobial compounds with compounds that stimulate various pathways of the immune system:

Many higher plants contain novel metabolites with antimicrobial and antiviral properties. However, in the developed world almost all clinically used chemotherapeutics have been produced by in vitro chemical synthesis.

Exceptions, like taxol and vincristine, were structurally complex metabolites that were difficult to synthesize in vitro. Many non-natural, synthetic drugs cause severe side effects that were not acceptable except as treatments of last resort for terminal diseases such as cancer.

The metabolites discovered in medicinal plants may avoid the side effect of synthetic drugs, because they must accumulate within living cells.

Semisynthetic procedures can sometimes get around these problems. This often involves harvesting a biosynthetic intermediate from the natural source, rather than the final (lead) compound itself.

The intermediate could then be converted to the final product by conventional synthesis. This approach can have two advantages. First, the intermediate may be more easily extracted in higher yield than the final product itself.

Second, it may allow the possibility of synthesizing analogues of the final product. The semisynthetic penicillins are an illustration of this approach. Another recent example is that of paclitaxel. It is manufactured by extracting 10-deacetylbaccatin III from the needles of the yew tree, then carrying out a four-stage synthesis.

Drugs

A number of drugs have been derived from biological a source in nature these include:.

- Amrubicin
- Anidulafungin
- Apomorphine
- Artemotil

- Aztreonam
- Biapenem
- Bivalirudin
- Bleomycin
- Capsaicin
- Caspofungin
- Cefditoren
- Codinaeopsin
- Colchicine
- Daptomycin
- Dimethyltryptamine
- Doripenem
- Dronabinol
- Ergotamine
- Ertapenem
- Everolimus
- Exenatide
- Fumagillin
- Galantamine
- Ixabepilone
- Lisdexamfetamine
- Mescaline
- Methylnaltrexone
- Micafungin
- Miglustat
- Mycophenolate
- Nitisinone
- Orlistat
- Paclitaxel
- Phenethylamine
- Pimecrolimus
- Retapamulin
- Romidepsin

- Rosuvastatin
- Spiruchostatins
- Telavancin
- Telithromycin
- Temsirolimus
- Tigecycline
- Tiotropium
- Trabectedin
- Ziconotide
- Zotarolimus

Phytotherapy

Phytotherapy is the study of the use of extracts from natural origin as medicines or health-promoting agents.

Traditional phytotherapy is a synonym for herbalism and regarded as alternative medicine by much of Western medicine.

Although the medicinal and biological effects of many plant constituents such as alkaloids (morphine, atropine etc.) have been proven through clinical studies, there is debate about the efficacy and the place of phytotherapy in medical therapies.

Modern phytotherapy, following the scientific method, can be considered the study on the effects and clinical use of herbal medicines.

Standardization

In herbal medicine, plant material that has been processed in a repeatable operation so that a discrete marker constituent is at a verified concentration is then considered standardized.

Active constituent concentrations may be misleading measures of potency if cofactors are not present.

A further problem is that the important constituent is often unknown. For instance St John's wort is often standardized to the antiviral constituent hypericin which is now known to be the active ingredient for antidepressant use. Other companies standardize to hyperforin or both, although there may be some 24 known possible constituents.

Only a minority of chemicals used as standardization markers are known to be active constituents. Standardization has not been standardized yet: different companies use different markers, or different levels of the same markers, or different methods of testing for marker compounds. Herbalist and manufacturer David Winston points out that whenever different compounds are chosen as 'active ingredients' for different herbs, there is a chance that suppliers will get a substandard batch (low on the chemical markers) and mix it with a batch higher in the desired marker to compensate for the difference.

Quality

The quality of crude drugs or plant medicines depends upon a variety of factors, including the variability in the specie (or species) of plant being used; the plant's growing conditions (i.e. soil, sun, climate); and the timing of harvest, post-harvest processing, and storage conditions. The quality of some plant drugs can judged by Organoleptic factors (i.e. sensory properties such as the taste, colour, odour or feel of the drug), or by administering a small dose of the drug and observing the effects. These conditions have been noted in historical herbals such as Culpepper's Complete Herbal or The Shen Nong or Divine Farmer's Materia Medica. This was standard pharmacognosy curriculum for many years. Storage after collection is a factor worthy of study; researchers in Nara, Japan have stored samples of ginseng root (Panax ginseng), licorice root (Glycyrrhiza glabra) and rhubarb root (Rheum emodi) that have been shown to retain their active properties for over 1,200 years. In modern times the foregoing aspects are no less important, but have been neglected with the advent of laboratory testing, although it generally is true that only certain constituents are identified and measured. Processes like HPLC (High performance liquid chromatography), GC (gas chromatography), UV/VIS (Ultraviolet/Visible spectrophotometry) or AA (atomic absorption spectroscopy) are used to identify species, measure bacteriological contamination, assess potency and eventually creating Certificates of Analysis for the material. Quality should be overseen by either authorities ensuring Good Manufacturing Practices or regulatory agencies by the US FDA. In the United States one frequently sees comments that herbal medicine is unregulated, but this is not correct since the FDA and GMP regulations are in place. In Germany, the Commission E has produced a book of German legal-medical regulations which includes quality standards.

Investigations by the Scientific Community

Common sense, as advocated above, is essential when working with, and consuming, plant-based medicinal products. Research by the scientific community attempts to explore, validate — and above all — understand the physiological aspects of traditional herbal and vegetative treatments.

Safety

A number of herbs are thought to be likely to cause adverse effects. Such herbs, like most herbs for example in Traditional Chinese Medicine, are used in small doses within the herbal formula. Furthermore, "adulteration, inappropriate formulation, or lack of understanding of plant and drug interactions have led to adverse reactions that are sometimes life threatening or lethal." Proper double-blind clinical trials are needed to determine the safety and efficacy of each plant before they can be recommended for medical use.

Although many consumers believe that herbal medicines are safe because they are "natural", herbal medicines may interact with synthetic drugs causing toxicity to the patient, may have contamination that is a safety consideration, and herbal medicines, without proven efficacy, may be used to replace medicines that have a proven efficacy.

The political issues around the safety of crude drugs vary from considering natural remedies "safe" regardless of potential dangers to considering them a dangerous unknown.

Ephedra has been known to have numerous side effects, including severe skin reactions, irritability, nervousness, dizziness, trembling, headache, insomnia, profuse perspiration, dehydration, itchy scalp and skin, vomiting, hyperthermia, irregular heartbeat, seizures, heart attack, stroke, or death.

Ephedra has been an object of difficulty; having legitimate western medicine uses, illegal uses and powerful side effects. Known and used as Mormon Tea or Indian Tea, the plant contains the potent chemical drugs ephedrine and pseudoephedrine.

Aside from being chemicals used to create methamphetamine they have direct CNS stimulant effects including high blood pressure and high heart rate. These effects have led to strokes and other CNS or cardiac issues in certain people at certain dosages.

In recent years, the safety of ephedra-containing dietary supplements has been questioned by the United States Food and Drug Administration (FDA), the National Centre for Complementary and Alternative Medicine, and the medical community as a result of reports of serious side effects and ephedra-related deaths.

However, when used appropriately by the correct people it is an effective decongestant, a bronchodilator for use in asthma and an adjuvant for the common cold.

There is no evidence to conclude that herbs have more side effects and adverse actions than western (chemically synthesized) medications, which routinely have the same adverse side effects declared on their packages.

Poisonous plants which have limited medicinal effects are often not sold in material doses in the United States or are available only to trained pract-itioners, these include:

- Aconite
- Arnica
- Belladonna
- Bryonia
- Datura
- Gelsemium
- Henbane
- Male Fern
- Phytolacca
- Podophyllum
- Veratrum

Other plants contain potent alkaloids which may cause physical harm when used incorrectly, but are not treated as possibly dangerous, leaving an uninformed public at risk for side effects or herb to chemical drug interactions.

Gingko biloba arguably has positive effects for many people but it is a blood thinner which may increase or cause spontaneous bleeding. White Willow, the source of salicin which through salicylic acid is the base for acetylsalicylic acid or Asprin.

Salicylate drugs such as A.S.A. are effective in reducing pain and fever but are also blood thinners. If, for example, one were to take Gingko Biloba and White Willow reduced blood clotting may lead to increased length of wound bleeding or to situations such as hematuria.

Plants such as Comfrey and Petasites have specific toxicity due to hepatotoxic pyrrolizidinc alkaloid content. There are other plant medicines which require caution or can interact with other medications, including St. John's wort and grapefruit.

"Paraherbalism"

Phytochemical researcher Varro Eugene Tyler described paraherbalism as "faulty or inferior herbalism based on pseudoscience", using scientific terminology but lacking scientific evidence for safety and efficacy.

Tyler listed ten fallacies that distinguished herbalism from paraherbalism, including claims that there is a conspiracy to suppress safe and effective herbs, herbs can not cause harm, that whole herbs are more effective than molecules isolated from the plants, herbs are superior to drugs, the doctrine of signatures (the belief that the shape of the plant indicates its function) is valid, dilution of substances increases their potency (a doctrine of the pseudoscience of homeopathy), astrological alignments are significant, animal testing is not appropriate to indicate human effects, anecdotal evidence is an effective means of proving a substance works and herbs were created by God to cure disease. None of these beliefs have any basis in fact.

Carbohydrates

A carbohydrate is an organic compound that consists only of carbon, hydrogen, and oxygen, usually with a hydrogen: oxygen atom ratio of 2:1 (as in water); in other words, with the empirical formula $C_m(H_2O)_n$. (Some exceptions exist; for example, deoxyribose, a component of DNA, has the empirical formula $C_5H_{10}O_4$.) Carbohydrates are not technically hydrates of carbon. Structurally it is more accurate to view them as polyhydroxy aldehydes and ketones. The term is most common in biochemistry, where it is a synonym of saccharide. The carbohydrates (saccharides) are divided into four chemical groupings: monosaccharides, disaccharides, oligosaccharides, and polysaccharides.

In general, the monosaccharides and disaccharides, which are smaller (lower molecular weight) carbohydrates, are commonly referred to as sugars. The word *saccharide* comes from the Greek word *σÜκχαñïí* (*sákkharon*), meaning "sugar". While the scientific nomenclature of carbohydrates is complex, the names of the monosaccharides and disaccharides very often end in the suffix -ose. For example, blood sugar is the monosaccharide glucose, table sugar is the disaccharide sucrose, and milk sugar is the disaccharide lactose.

Carbohydrates perform numerous roles in living organisms. Polysaccharides serve for the storage of energy (e.g., starch and glycogen), and as structural components (e.g., cellulose in plants and chitin in arthropods).

The 5-carbon monosaccharide ribose is an important component of coenzymes (e.g., ATP, FAD, and NAD) and the backbone of the genetic molecule known as RNA. The related deoxyribose is a component of DNA. Saccharides and their derivatives include many other important biomolecules that play key roles in the immune system, fertilization, preventing pathogenesis, blood clotting, and development.

In food science and in many informal contexts, the term carbohydrate often means any food that is particularly rich in the complex carbohydrate starch (such as cereals, bread, and pasta) or simple carbohydrates, such as sugar (found in candy, jams, and desserts).

Formerly the name "carbohydrate" was used in chemistry for any compound with the formula $C_m(H_2O)_n$. Following this definition, some chemists considered formaldehyde (CH_2O) to be the simplest carbohydrate, while others claimed that title for glycolaldehyde. Today the term is generally understood in the biochemistry sense, which excludes compounds with only one or two carbons.

Natural saccharides are generally built of simple carbohydrates called monosaccharides with general formula

$(CH_2O)_n$ where n is three or more. A typical monosaccharide has the structure H-$(CHOH)_x$(C=O)-$(CHOH)_y$-H, that is, an aldehyde or ketone with many hydroxyl groups added, usually one on each carbon atom that is not part of the aldehyde or ketone functional group.

Examples of monosaccharides are glucose, fructose, and glyceraldehydes. However, some biological substances commonly called "monosaccharides" do not conform to this formula (e.g., uronic acids and deoxy-sugars such as fucose), and there are many chemicals that do conform to this formula but are not considered to be monosaccharides (e.g., formaldehyde CH_2O and inositol $(CH_2O)_6$).

The open-chain form of a monosaccharide often coexists with a closed ring form where the aldehyde/ketone carbonyl group carbon (C=O) and hydroxyl group (-OH) react forming a hemiacetal with a new C-O-C bridge.

Monosaccharides can be linked together into what are called polysaccharides (or oligosaccharides) in a large variety of ways. Many carbohydrates contain one or more modified monosaccharide units that have had one or more groups replaced or removed. For example, deoxyribose, a component of DNA, is a modified version of ribose; chitin is composed of repeating units of N-acetyl glucosamine, a nitrogen-containing form of glucose.

Formation

Disaccharides are formed when two monosaccharides are joined together and a molecule of water is removed, a process known as condensation. For example; milk sugar (lactose) is made from glucose and galactose whereas the sugar from sugar cane and sugar beets (sucrose) is made from glucose and fructose.

Maltose, another notable disaccharide, is made up of two glucose molecules. The two monosaccharides are bonded via a dehydration reaction (also called a condensation reaction or dehydration synthesis) that leads to the loss of a molecule of water and formation of a glycosidic bond.

Nutrition

Foods high in carbohydrate include fruits, sweets, soft drinks, breads, pastas, beans, potatoes, bran, rice, and cereals.

Carbohydrates are a common source of energy in living organisms; however, no carbohydrate is an essential nutrient in humans. Carbohydrates are not necessary building blocks of other molecules, and the body can obtain all its energy from protein and fats. The brain and neurons generally cannot burn fat for energy, but use glucose or ketones.

Humans can synthesize some glucose (in a set of processes known as gluconeogenesis) from specific amino acids, from the glycerol backbone in triglycerides and in some cases from fatty acids. Carbohydrate and protein contain 4 kilocalories per gram, while fats contain 9 kilocalories per gram. In the case of protein, this is somewhat misleading as only some amino acids are usable for fuel.

Organisms typically cannot metabolize all types of carbohydrate to yield energy. Glucose is a nearly universal and accessible source of calories. Many organisms also have the ability to metabolize other monosaccharides and Disaccharides, though glucose is preferred. In *Escherichia coli*, for example, the *lac* operon will express enzymes for the digestion of lactose when it is present, but if both lactose and glucose are present the *lac* operon is repressed, resulting in the glucose being used first. Polysaccharides are also common sources of energy. Many organisms can easily break down starches into glucose, however, most organisms cannot metabolize cellulose or other polysaccharides like chitin and arabinoxylans. These carbohydrates types can be metabolized by some bacteria and protists. Ruminants and termites, for example, use microorganisms to process cellulose. Even though these complex carbohydrates are not very digestible, they may comprise important dietary elements for humans. Called dietary fibre, these carbohydrates enhance digestion among other benefits.

Based on the effects on risk of heart disease and obesity, the Institute of Medicine recommends that American and Canadian adults get between 45–65% of dietary energy from carbohydrates. The Food and Agriculture Organisation and World Health Organisation jointly recommend that national dietary guidelines set a goal of 55–75% of total energy from carbohydrates, but only 10% directly from sugars (their term for simple carbohydrates).

Classification

Historically nutritionists have classified carbohydrates as either simple or complex. However, the exact delineation of these categories is ambiguous. Today, simple carbohydrate typically refers to monosaccharides and disaccharides and complex carbohydrate means polysaccharides (and oligosaccharides). However, the term *complex carbohydrate* was first used in slightly different context in the U.S. Senate Select Committee on Nutrition and Human Needs publication *Dietary Goals for the United States* (1977).

In this work, complex carbohydrate were defined as "fruit, vegetables and whole-grains". Some nutritionists use complex carbohydrate to refer to any sort of digestible saccharide present in a whole food, where fibre, vitamins and minerals are also found (as opposed to processed carbohydrates, which provide calories but few other nutrients).

Some simple carbohydrates (e.g. fructose) are digested very slowly, while some complex carbohydrates (starches), especially if processed, raise blood sugar rapidly. The speed of digestion is determined by a variety of factors including which other nutrients are consumed with the carbohydrate, how the food is prepared, individual differences in metabolism, and the chemistry of the carbohydrate.

The USDA's *Dietary Guidelines for Americans 2010* call for moderate- to high-carbohydrate consumption from a balanced diet that includes six one-ounce servings of grain foods each day, at least half from whole grain sources and the rest from enriched.

The glycemic index (GI) and glycemic load concepts have been developed to characterise food behaviour during human digestion.

They rank carbohydrate-rich foods based on the rapidity and magnitude of their effect on blood glucose levels. Glycemic index is a measure of how quickly food glucose is absorbed, while glycemic load is a measure of the total absorbable glucose in foods. The insulin index is a similar, more recent classification method that ranks foods based on their effects on blood insulin levels, which are caused by glucose (or starch) and some amino acids in food.

Metabolism

Carbohydrate metabolism denotes the various biochemical processes responsible for the formation, breakdown and interconversion of carbohydrates in living organisms.

The most important carbohydrate is glucose, a simple sugar (monosaccharide) that is metabolized by nearly all known organisms. Glucose and other carbohydrates are part of a wide variety of metabolic pathways across species: plants synthesize carbohydrates from atmospheric gases by photosynthesis storing the absorbed energy internally, often in the form of starch or lipids. Plant components are eaten by animals and fungi, and used as fuel for cellular respiration. Oxidation of one gram of carbohydrate yields approximately 4 kcal of energy and from lipids about 9 kcal. Energy obtained from metabolism (e.g. oxidation of glucose) is usually stored temporarily within cells in the form of ATP. Organisms capable of aerobic respiration metabolize glucose and oxygen to release energy with carbon dioxide and water as byproducts. Carbohydrates include complex and simple sugars. Simple sugars can be broken down directly in cells. Complex carbohydrates such as sucrose (a disaccharide, or a molecule containing two simple sugars) contain more than one simple sugar in a chain. They are broken down in the gut by specific enzymes that break the chain and release simple sugars. Starch is a polymer of glucose units and is broken down to glucose. Cellulose is a carbohydrate chain that cannot be digested by some animals. Some bacteria that produce enzymes for cellulose live inside the gut of some mammals such as cows, and when cows eat grass, the cellulose is broken down by the bacteria and some of it is released into the gut. Carbohydrates are a superior short-term fuel for organisms because they are simpler to metabolize than fats or those amino acid portions of proteins that are used for fuel. In animals, the most important carbohydrate is glucose; the level of glucose is used as the main control for the central metabolic hormone, insulin. Starch, and cellulose in a few organisms (e.g., termites, ruminants, and some bacteria), both being glucose polymers, are disassembled during digestion and absorbed as glucose. Some simple carbohydrates have their own enzymatic oxidation pathways, as do only a few of the more complex carbohydrates. The disaccharide lactose, for instance, requires the enzyme lactase to be broken into its monosaccharides components; many animals lack this enzyme in adulthood. Carbohydrates are typically stored as long polymers of glucose molecules with glycosidic bonds for structural support (e.g. chitin, cellulose) or for energy storage (e.g. glycogen, starch). However, the strong affinity of most carbohydrates for water makes storage of large quantities of carbohydrates inefficient due to the large molecular weight of the solvated water-carbohydrate complex.

In most organisms, excess carbohydrates are regularly catabolised to form acetyl-CoA, which is a feed stock for the fatty acid synthesis pathway; fatty acids, triglycerides, and other lipids are commonly used for long-term energy storage. The hydrophobic character of lipids makes them a much more compact form of energy storage than hydrophilic carbohydrates. However, animals, including humans, lack the necessary enzymatic machinery and so do not synthesize glucose from lipids, though glycerol can be converted to glucose. All carbohydrates share a general formula of approximately $C_nH_{2n}O_n$; glucose is $C_6H_{12}O_6$. Monosaccharides may be chemically bonded together to form disaccharides such as sucrose and longer polysaccharides such as starch and cellulose. Oligo/ polysaccharides are typically cleaved into smaller monosaccharides by enzymes called glycoside hydrolases. The monosaccharide units then enter monosaccharide catabolism. Organisms vary in the range of monosaccharides they can absorb and use, and also in the range of more complex carbohydrates they are capable of disassembling.

Metabolic pathways:

- Carbon fixation, or photosynthesis, in which CO_2 is reduced to carbohydrate.
- Glycolysis - the oxidation metabolism of glucose molecules to obtain ATP and pyruvate
  - o Pyruvate from glycolysis enters the Krebs cycle, also known as the citric acid cycle, in aerobic organisms after moving through pyruvate dehydrogenase complex.
- The pentose phosphate pathway, which acts in the conversion of hexoses into pentoses and in NADPH regeneration.

- Glycogenesis - the conversion of excess glucose into glycogen as a cellular storage mechanism; this prevents excessive osmotic pressure buildup inside the cell
- Glycogenolysis - the breakdown of glycogen into glucose, which provides a glucose supply for glucose-dependent tissues.
- Gluconeogenesis - *de novo* synthesis of glucose molecules from simple organic compounds. An example in humans is the conversion of a few amino acids in cellular protein to glucose. Metabolic use of glucose is highly important as an energy source for muscle cells and in the brain, and red blood cells.

Glucoregulation

Glucoregulation is the maintenance of steady levels of glucose in the body; it is part of homeostasis, and so keeps a constant internal environment around cells in the body. The hormone insulin is the primary regulatory signal in animals, suggesting that the basic mechanism is very old and very central to animal life. When present, it causes many tissue cells to take up glucose from the circulation, causes some cells to store glucose internally in the form of glycogen, causes some cells to take in and hold lipids, and in many cases controls cellular electrolyte balances and amino acid uptake as well. Its absence turns off glucose uptake into cells, reverses electrolyte adjustments, begins glycogen breakdown and glucose release into the circulation by some cells, begins lipid release *from* lipid storage cells, etc. The level of circulatory glucose (known informally as "blood sugar") is the most important signal to the insulin-producing cells. Because the level of circulatory glucose is largely determined by the intake of dietary carbohydrates, diet controls major aspects of metabolism via insulin. In humans, insulin is made by beta cells in the pancreas, fat is stored in adipose tissue cells, and glycogen is both stored and released as needed by liver cells. Regardless of insulin levels, no glucose is released to the blood from internal glycogen stores from muscle cells.

The hormone glucagon, on the other hand, has an effect opposite to that of insulin, forcing the conversion of glycogen in liver cells to glucose, which is then released into the blood.

Muscle cells, however, lack the ability to export glucose into the blood. The release of glucagon is precipitated by low levels of blood glucose. Other hormones, notably growth hormone, cortisol, and certain catecholamines (such as epinepherine) have glucoregulatory actions similar to glucagon.

Catabolism

Catabolism is the metabolic reaction in which cells undergo to extract energy. There are two major metabolic pathways of monosaccharide catabolism: glycolysis and the citric acid cycle. In glycolysis, oligo/polysaccharides are cleaved first to smaller monosaccharides by enzymes called glycoside hydrolases. The monosaccharide units can then enter into monosaccharide catabolism. In some cases, as with humans, not all carbohydrate types are usable as the digestive and metabolic enzymes necessary are not present.

Carbohydrate Chemistry

Carbohydrate chemistry is a large and economically important branch of organic chemistry. Some of the main organic reactions that involve carbohydrates are:

- Carbohydrate acetalisation
- Cyanohydrin reaction
- Lobry-de Bruyn-van Ekenstein transformation
- Amadori rearrangement
- Nef reaction
- Wohl degradation
- Koenigs–Knorr reaction

Lipids

Lipids constitute a broad group of naturally occurring molecules that include fats, waxes, sterols, fat-soluble vitamins (such as vitamins A, D, E, and K), monoglycerides, diglycerides, triglycerides, phospholipids, and others. The main biological functions of lipids include energy storage, as structural components of cell membranes, and as important signaling molecules.

Lipids may be broadly defined as hydrophobic or amphiphilic small molecules; the amphiphilic nature of some lipids allows them to form structures such as vesicles, liposomes, or membranes in an aqueous environment. Biological lipids originate entirely or in part from two distinct types of biochemical subunits or "building-blocks": ketoacyl and isoprene groups. Using this approach, lipids may be divided into eight categories: fatty acids, glycerolipids, glycerophospholipids, sphingolipids, saccharolipids, and polyketides (derived from condensation of ketoacyl subunits); and sterol lipids and prenol lipids (derived from condensation of isoprene subunits). Although the term *lipid* is sometimes used as a synonym for fats, fats are a subgroup of lipids called triglycerides. Lipids also encompass molecules such as fatty acids and their derivatives (including tri-, di-, monoglycerides, and phospholipids), as well as other sterol-containing metabolites such as cholesterol. Although humans and other mammals use various biosynthetic pathways to both break down and synthesize lipids, some essential lipids cannot be made this way and must be obtained from the diet.

Fatty Acids

Fatty acids, or fatty acid residues when they form part of a lipid, are a diverse group of molecules synthesized by chain-elongation of an acetyl-CoA primer with malonyl-CoA or methylmalonyl-CoA groups in a process called fatty acid synthesis. They are made of a hydrocarbon chain that terminates with a carboxylic acid group; this arrangement confers the molecule with a polar, hydrophilic end, and a nonpolar, hydrophobic end that is insoluble in water. The fatty acid structure is one of the most fundamental categories of biological lipids, and is commonly used as a building-block of more structurally complex lipids. The carbon chain, typically between four and 24 carbons long, may be saturated or unsaturated, and may be attached to functional groups containing oxygen, halogens, nitrogen, and sulfur. Where a double bond exists, there is the possibility of either a *cis* or a *trans* geometric isomerism, which significantly affects the molecule's molecular configuration. *Cis*-double bonds cause the fatty acid chain to bend, an effect that is more pronounced the more double bonds there are in a chain. This in turn plays an important role in the structure and function of cell membranes. Most naturally occurring fatty acids are of the *cis* configuration, although the *trans* form does exist in some natural and partially hydrogenated fats and oils. Examples of biologically important fatty acids are the eicosanoids, derived primarily from arachidonic acid and eicosapentaenoic acid, that include prostaglandins, leukotrienes, and thromboxanes. Docosahexaenoic acid is also important in biological systems, particularly with respect to sight. Other major lipid classes in the fatty acid category are the fatty esters and fatty amides. Fatty esters include important biochemical intermediates such as wax esters, fatty acid thioester coenzyme A derivatives, fatty acid thioester ACP derivatives and fatty acid carnitines. The fatty amides include N-acyl ethanolamines, such as the cannabinoid neurotransmitter anandamide.

Glycerolipids

Glycerolipids are composed mainly of mono-, di-, and tri-substituted glycerols, the most well-known being the fatty acid triesters of glycerol, called triglycerides. The word triacylglycerol is sometimes used synonymously with triglyceride, however this is misleading with respect to these compounds as they contain no hydroxyl group. In these compounds, the three hydroxyl groups of glycerol are each esterified, typically by different fatty acids. Because they function as an energy store, these lipids comprise the bulk of storage fat in animal tissues. The hydrolysis of the ester bonds of triglycerides and the release of glycerol and fatty acids from adipose tissue are the initial steps in metabolising fat. Additional subclasses of glycerolipids are represented by glycosylglycerols, which are characterised by the presence of one or more sugar residues attached to glycerol via a glycosidic linkage. Examples of structures in this category are the digalactosyldiacylglycerols found in plant membranes and seminolipid from mammalian sperm cells.

Sphingolipids

Sphingolipids are a complicated family of compounds that share a common structural feature, a sphingoid base backbone that is synthesized *de novo* from the amino acid serine and a long-chain fatty acyl CoA, then converted into ceramides, phosphosphingolipids, glycosphingolipids and other compounds. The major sphingoid base of mammals is commonly referred to as sphingosine. Ceramides (N-acyl-sphingoid bases) are a major subclass of sphingoid base derivatives with an amide-linked fatty acid. The fatty acids are typically saturated or mono-unsaturated with chain lengths from 16 to 26 carbon atoms. The major phosphosphingolipids of mammals are sphingomyelins (ceramide phosphocholines), whereas insects contain mainly ceramide phosphoethanolami-nes and fungi have phytoceramide phosphoinositols and mannose-containing headgroups. The glycosphingolipids are a diverse family of molecules composed of one or more sugar residues linked via a glycosidic bond to the sphingoid base. Examples of these are the simple and complex glycosphingolipids such as cerebrosides and gangliosides.

Sterol Lipids

Sterol lipids, such as cholesterol and its derivatives, are an important component of membrane lipids, along with the glycerophospholipids and sphingomyelins. The steroids, all derived from the same fused four-ring core structure, have different biological roles as hormones and signaling molecules. The eighteen-carbon (C18) steroids include the estrogen family whereas the C19 steroids comprise the androgens such as testosterone and androsterone. The C21 subclass includes the progestogens as well as the glucocorticoids and mineralocorticoids. The secosteroids, comprising various forms of vitamin D, are characterised by cleavage of the B ring of the core structure. Other examples of sterols are the bile acids and their conjugates, which in mammals are oxidized derivatives of cholesterol and are synthesized in the liver. The plant equivalents are the phytosterols, such as β-sitosterol, stigmasterol, and brassicasterol; the latter compound is also used as a biomarker for algal growth. The predominant sterol in fungal cell membranes is ergosterol.

Chapter 8

Thermal Equilibrium

Thermal equilibrium is achieved when two systems in thermal contact with each other cease to have a net exchange of energy. It follows that if two systems are in thermal equilibrium, then their temperatures are the same.

Thermal equilibrium occurs when a system's macroscopic thermal observables have ceased to change with time. For example, an ideal gas whose distribution function has stabilised to a specific Maxwell-Boltzmann distribution would be in thermal equilibrium. This outcome allows a single temperature and pressure to be attributed to the whole system. Thermal equilibrium of a system does not imply absolute uniformity within a system; for example, a river system can be in thermal equilibrium when the macroscopic temperature distribution is stable and not changing in time, even though the spatial temperature distribution reflects thermal pollution inputs.

Thermal equilibrium is a theoretical physical concept, used especially in theoretical texts, that means that all temperatures of interest are unchanging in time and uniform in space. When the temperatures of interest are just those in the different parts of one body, the concept also requires that any flow of heat by thermal conduction or by thermal radiation into or out of one part of the body be balanced by a flow of heat in the opposite sense into or out of another part of the body. When the temperatures of interest belong to several bodies, the concept also requires that flows of heat between each pair of bodies balance to a zero net flow, but it allows the several bodies to gain or lose heat to several external reservoirs provided that their total rate of inflow from all reservoirs is equal to their total rate of outflow to all reservoirs and that each flow is unchanging in time. For some situations, the definition of transfer of heat can be problematic.

Some writers use the term thermal equilibrium in a different sense. They mean by it that the spatial temperature distribution of the body is not necessarily uniform, and indeed is likely to be non-uniform, but is maintained unvarying in time, by flows of energy; for example they mean that there is spatially distributed radiative cooling of the body and equal and opposite spatially distributed energy addition by condensation of water vapour, just so as on average to keep the spatial distribution of temperature time-invariant. Thermal equilibrium does not mean the same as thermodynamic equilibrium, because the latter requires that there be equilibrium of all kinds, not only thermal, and that there be no flow of any kind, in the system of interest.

Transfer of Heat by Conduction and Radiation between Systems

Heat can flow into or out of a closed system by way of thermal conduction or of thermal radiation to or from a thermal reservoir, and when this process is effecting net transfer of heat, the system's temperature can be changing, and if so, the system is not in thermal equilibrium.

For thermodynamics, it is considered that convection transports internal energy by bulk flow; this is considered not to be transfer of heat as such.

Change of Internal State of an Isolated System

If an isolated system is left long enough, it will reach a state of thermal equilibrium, in which its temperature will be uniform throughout, but not necessarily a state of thermodynamic equilibrium, if there is some structural barrier that can prevent some possible processes in the system from reaching equilibrium. An isolated system can change its temperature or its spatial distribution of temperature by changing the state of its materials. A rod of iron, initially prepared to be hot at one end and cold at the other, when isolated, will change so that its temperature becomes uniform all along its length; during the process, the rod is not in thermal equilibrium until its temperature is uniform. A very tall isolating vessel initially containing a thermally heterogeneous distribution of material, left for a long time under the influence of a steady gravitational field, along its tall dimension, due to an outside body, will settle to a state of uniform temperature though not of uniform pressure or density, and is then in thermal equilibrium and even of thermodynamic equilibrium. A system prepared as a mixture of petrol vapour and air can be ignited by a spark and produce carbon dioxide and water; if this happens in an isolated system, it will increase the temperature of the system, and during the increase, the system is not in thermal equilibrium; but eventually the system will settle to a uniform temperature.

In a system prepared as a block of ice floating in a bath of hot water, and then isolated, the ice can melt; during the melting, the system is not in thermal equilibrium; but eventually its temperature will become uniform. Such changes in isolated systems are irreversible in the sense that while such a change will occur spontaneously whenever the system is prepared in the same way, the reverse change will never occur spontanteously within the isolated system; this is a large part of the content of the second law of thermodynamics. Truly isolated systems hardly occur in nature, and nearly always are artificially prepared.

Bodies prepared with separately uniform temperatures, then put into purely thermal communication with each other. If bodies are prepared with separately uniform temperatures, and are then put into purely thermal communication with each other, by conductive or radiative pathways, they will be in thermal equilibrium with each other just when they have the same temperature, and then there will be no net transfer of heat between them; but if initially they do not have the same temperature, heat will flow from the hotter to the colder, by whatever pathway, conductive or radiative, is available, and this flow will continue until thermal equilibrium is reached and then they will have the same temperature. One form of thermal equilibrium is radiative exchange equilibrium. Two bodies, each with its own uniform temperature, in solely radiative connection, no matter how far apart, or what partially obstructive, reflective, or refractive, obstacles lie in their path of radiative exchange, not moving relative to one another, will exchange thermal radiation, in net the hotter transferring energy to the cooler, and will exchange equal and opposite amounts just when they are at the same temperature. In this situation, Kirchhoff's law of equality of radiative emissivity and absorptivity and the Helmholtz reciprocity principle are in play.

Distinctions between Thermal and Thermodynamic Equilibria

In the seventh German edition of his *Treatise on Thermodynamics*, Planck mostly writes of 'thermal equilibrium' , and only sparingly of 'thermodynamical equilibrium' . Though Planck does not explicitly say so, it seems from the context of his text that he may mean that thermal equilibrium is a good concept for cases in which chemical changes are not allowed, while thermodynamic equilibrium is a good concept when chemical changes are allowed. In his *The Theory of Heat Radiation*, Planck freely writes of 'thermodynamic equilibrium', but not at all of 'thermal equilibrium'; in this case, radiative transfer of heat is being considered, while that is not mentioned in the *Treatise on Thermodynamics*.

An explicit distinction between 'thermal equilibrium' and 'thermodynamic equilibrium' is made by C.J. Adkins. He allows that two systems might be allowed to exchange heat but be constrained from exchanging work; they will naturally exchange heat till they have equal temperatures, and reach thermal equilibrium, but in general will not be in thermodynamic equilibrium. They can reach thermodynamic equilibrium when they are allowed also to exchange work. Another explicit distinction between 'thermal equilibrium' and 'thermodynamic equilibrium' is made by B. C. Eu. He considers two systems in thermal contact, one a thermometre, the other a system in which several irreversible processes are occurring. He considers the case in which, over the time scale of interest, it happens that both the thermometre reading and the irreversible processes are steady. Then there is thermal equilibrium without thermodynamic equilibrium. Eu proposes consequently that the zeroth law of thermodynamics can be considered to apply even when thermodynamic equilibrium is not present; also he proposes that if changes are occurring so fast that a steady temperature cannot be defined, then "it is no longer possible to describe the process by means of a thermodynamic formalism. In other words, thermodynamics has no meaning for such a process."

Theoretical Foundations

It is scientifically permissible, and perhaps unavoidably necessary, to start a project of reasoning with several mutually coherent and dependent primitive presupposed concepts. The concept of thermal equilibrium is thus coherent with the concepts of temperature and of heat transfer.

These three concepts hardly make physical sense without each other. They were considered coordinately before and during and after the developments of calorimetry and of thermodynamics, for example by Maxwell and by Planck.

Instead of relying on this triple of jointly defined physical concepts, some writers, motivated by a desire for axiomatic parsimony and precision or mathematical elegance, prefer to define thermal equilibrium by relying on a presupposed notion of thermodynamic equilibrium, in which all mechanically measurable properties of a body have become stationary, and one infers that consequently the otherwise undefined thermal properties also are stationary. Carathéodory is an example of such writers, as seen in his 1909 article. This approach leaves one at the mercy of the questions of what is meant physically by "all mechanically measurable properties" and what is meant by saying that they "have become stationary", and still relying on some concept that allows bodies not 'thermally connected' to be put into 'thermal connection'.

It also leaves a person, who does not know in advance the notions of heat transfer and temperature, reliant on the assumptions, such as conservation of energy, and on the mathematical development, of the theory of thermodynamics.

Quasistatic Equilibrium

Quasistatic equilibrium is the quasi-balanced state of a thermodynamic system near to thermodynamic equilibrium, in some sense. In a quasistatic or equilibrium process, a sufficiently slow transition of a thermodynamic system from one equilibrium state to another occurs such that at every moment in time the state of the system is close to an equilibrium state. During a quasistatic process, the system reaches equilibrium much faster, almost instantaneously, than its physical parameters vary.

In thermodynamics, a quasistatic process is a thermodynamic process that happens infinitely slowly. However, it is very important of note that *no* real process is quasistatic. Therefore in practice, such processes can only be approximated by performing them *infinitesimally* slowly.

Quasistatic and Reversibility

A quasistatic process ensures that the system will go through a sequence of states that are infinitesimally close to equilibrium (so the system remains in quasistatic equilibrium), in which case the process is typically reversible. Any reversible process is necessarily a quasistatic one.

However, some quasistatic processes are irreversible, if there is heat flowing (in or out of the system) or if entropy is being created in some other way. An example of a quasistatic process that is not reversible is a compression against a system with a piston subject to friction — Although the system is always in thermal equilibrium, the friction ensures the generation of dissipative entropy, which directly goes against the definition of reversible.

A notable example of a process that is not even quasistatic is the slow heat exchange between two bodies at two finitely different temperatures, where the heat exchange rate is controlled by an approximately adiabatic partition between the two bodies (Sears and Salinger, 1986) — in this case, no matter how slowly the process takes place, the states of the two bodies are never infinitesimally close to equilibrium, since thermal equilibrium requires that the two bodies be at precisely the same temperature. Some ambiguity exists in the literature concerning the distinction between quasistatic and reversible processes, as these are sometimes taken as synonyms (Bernard H Lavenda, 1978). The reason is precisely because of the proven theorem that any reversible process is also a quasistatic one, even though we have also shown that the converse is not true. It is practically not useful to differentiate between the two because any engineer would remember to include friction when calculating the dissipative entropy generation.

Non-Equilibrium

Non-equilibrium thermodynamics is a branch of thermodynamics that deals with systems that are not in thermodynamic equilibrium. Most systems found in nature are not in thermodynamic equilibrium because they are changing or can be triggered to change over time, and are continuously and discontinuously subject to flux of matter and energy to and from other systems. The thermodynamic study of non-equilibrium systems requires more general concepts than are dealt with by equilibrium thermodynamics. Many natural systems still today remain beyond the scope of currently known macroscopic thermodynamic methods. Non-equilibrium thermodynamics is a branch of thermodynamics that deals with systems that are not in thermodynamic equilibrium. Most systems found in nature are not in thermodynamic equilibrium; for they are changing or can be triggered to change over time, and are continuously and discontinuously subject to flux of matter and energy to and from other systems and to chemical reactions. Non-equilibrium thermodynamics is concerned with transport processes and with the rates of chemical reactions. Many natural systems still today remain beyond the scope of currently known macroscopic thermodynamic methods. The thermodynamic study of non-equilibrium systems requires more general concepts than are dealt with by equilibrium thermodynamics. One fundamental difference between equilibrium thermodynamics and non-equilibrium thermodynamics lies in the behaviour of inhomogeneous systems, which require for their study knowledge of rates of reaction which are not considered in equilibrium thermodynamics of homogeneous systems. This is discussed below. Another fundamental difference is the difficulty in defining entropy in macroscopic terms for systems not in thermodynamic equilibrium.

Non-equilibrium thermodynamics is a work in progress, not an established edifice. This chapter cannot give a full account of it. This chapter will try to sketch some approaches to it and some concepts important for it. Some concepts of particular importance for non-equilibrium thermodynamics include time rate of dissipation of energy (Rayleigh 1873, Onsager 1931, also), time rate of entropy production (Onsager 1931), thermodynamic fields, dissipative structure, and non-linear dynamical structure.

Of interest is the thermodynamic study of non-equilibrium steady states, in which entropy production and some flows are non-zero, but there is no time variation. One initial approach to non-equilibrium thermodynamics is sometimes called 'classical irreversible thermodynamics'. There are other approaches to non-equilibrium thermodynamics, for example 'extended irreversible thermodynamics', but they are hardly touched on in the present chapter.

Quasi-Radiationless Non-Equilibrium Thermodynamics of Matter in Laboratory Conditions

According to Wildt, current versions of non-equilibrium thermodynamics ignore radiant heat; they can do so because they refer to laboratory quantities of matter under laboratory conditions with temperatures well below those of stars. At laboratory temperatures, in laboratory quantities of matter, thermal radiation is weak and can be practically nearly ignored. For example, atmospheric physics is concerned with large amounts of matter, occupying cubic kilometres, that are not necessarily within the scope of laboratory conditions.

'Classical Irreversible Thermodynamics'

The term 'classical irreversible thermodynamics' is sometimes used to refer to a version of non-equilibrium thermodynamics that demands certain simplifying assumptions, as follows. The assumptions have the effect of making each very small volume element of the system effectively homogeneous, or well-mixed, or without an effective spatial structure, and without kinetic energy of bulk flow or of diffusive flux. Even within the thought-frame of classical irreversible thermodynamics, care is needed in choosing the independent variables for systems.

In some writings, it is assumed that the intensive variables of equilibrium thermodynamics are sufficient as the independent variables for the task (such variables are considered to have no 'memory', and do not show hysteresis); in particular, local flow intensive variables are not admitted as independent variables; local flows are considered as dependent on quasistatic local intensive variables. (In other writings, local flow variables are considered; examples with flows are in the thermoelectric phenomena known as the Seebeck and the Peltier effects, considered by Kelvin in the nineteenth century and by Onsager in the twentieth.

These effects occur at metal junctions, which were originally effectively treated as two-dimensional surfaces, with no spatial volume, and no spatial variation.) Also it is assumed that the local entropy density is the same function of the other local intensive variables as in equilibrium; this is called the local thermodynamic equilibrium assumption). In the classical irreversible thermodynamical approach, it is usually assumed that there is no bulk flow of ponderable matter, so that kinetic energy does not need to be considered.

Radiation is also ignored because it is transfer of energy between regions, which can be remote from one another. In the classical irreversible thermodynamic approach, there is allowed very small spatial variation, from very small volume element to adjacent very small volume element, but it is assumed that the global entropy of the system can be found by simple spatial integration of the local entropy density; this means that spatial structure cannot contribute as it properly should to the global entropy assessment for the system.

This approach assumes spatial and temporal continuity and even differentiability of locally defined intensive variables such as temperature and internal energy density. All of these are very stringent demands. Consequently, this approach can deal with only a very limited range of phenomena. This approach is nevertheless valuable as an introduction to non-equilibrium thermodynamics.

Basic Concepts

There are many examples of stationary non-equilibrium systems, some very simple, like a system confined between two thermostats at different temperatures or the ordinary Couette flow, a fluid enclosed between two flat walls moving in opposite directions and defining non-equilibrium conditions at the walls. Laser action is also a non-equilibrium process, but it depends on departure from local thermodynamic equilibrium and is thus beyond the scope of classical irreversible thermodynamics;

here a strong temperature difference is maintained between two molecular degrees of freedom (with molecular laser, vibrational and rotational molecular motion), the requirement for two component 'temperatures' in the one small region of space, precluding local thermodynamic equilibrium, which demands that only one temperature be needed. Damping of acoustical perturbations or shock waves are non-stationary non-equilibrium processes. Driven complex fluids, turbulent systems and glasses are other examples of non-equilibrium systems.

The mechanics of macroscopic systems depends on a number of extensive quantities. It should be stressed that all systems are permanently interacting with their surroundings, thereby causing unavoidable fluctuations of extensive quantities. Equilibrium conditions of thermodynamic systems are related to the maximum property of the entropy. If the only extensive quantity that is allowed to fluctuate is the internal energy, all the other ones being kept strictly constant, the temperature of the system is measurable and meaningful. The system's properties are then most conveniently described using the thermodynamic potential Helmholtz free energy ($A = U - TS$), a Legendre transformation of the energy. If, next to fluctuations of the energy, the macroscopic dimensions (volume) of the system are left fluctuating, we use the Gibbs free energy ($G = U + PV - TS$), where the system's properties are determined both by the temperature and by the pressure. Non-equilibrium systems are much more complex and they may undergo fluctuations of more extensive quantities. The boundary conditions impose on them particular intensive variables, like temperature gradients or distorted collective motions (shear motions, vortices, etc.), often called thermodynamic forces. If free energies are very useful in equilibrium thermodynamics, it must be stressed that there is no general law defining stationary non-equilibrium properties of the energy as is the second law of thermodynamics for the entropy in equilibrium thermodynamics.

The independent variables are the intensities. Intensities are global values, valid for the system as a whole. When boundaries impose to the system different local conditions, (e.g. temperature differences), there are intensive variables representing the average value and others representing gradients or higher moments. The latter are the thermodynamic forces driving fluxes of extensive properties through the system.

It may be shown that the Legendre transformation changes the maximum condition of the entropy (valid at equilibrium) in a minimum condition of the extended Massieu function for stationary states, no matter whether at equilibrium or not.

Stationary States, Fluctuations, and Stability

In thermodynamics one is often interested in a stationary state of a process, allowing that the stationary state include the occurrence of unpredictable and experimentally unreproducible fluctuations in the state of the system. The fluctuations are due to the system's internal sub-processes and to exchange of matter or energy with the system's surroundings that create the constraints that define the process.

If the stationary state of the process is stable, then the unreproducible fluctuations involve local transient decreases of entropy. The reproducible response of the system is then to increase the entropy back to its maximum by irreversible processes: the fluctuation cannot be reproduced with a significant level of probability. Fluctuations about stable stationary states are extremely small except near critical points (Kondepudi and Prigogine 1998, page 323). The stable stationary state has a local maximum of entropy and is locally the most reproducible state of the system. There are theorems about the irreversible dissipation of fluctuations. Here 'local' means local with respect to the abstract space of thermodynamic coordinates of state of the system.

If the stationary state is unstable, then any fluctuation will almost surely trigger the virtually explosive departure of the system from the unstable stationary state. This can be accompanied by increased export of entropy.

Local Thermodynamic Equilibrium

The scope of present-day non-equilibrium thermodynamics does not cover all physical processes. A condition for the validity of many studies in non-equilibrium thermodynamics of matter is that they deal with what is known as *local thermodynamic equilibrium*.

Local Thermodynamic Equilibrium of Ponderable Matter

Local thermodynamic equilibrium of matter means that conceptually, for study and analysis, the system can be spatially and temporally divided into 'cells' or 'micro-phases' of small (infinitesimal) size, in which classical thermodynamical equilibrium conditions for matter are fulfilled to good approximation. These conditions are unfulfilled, for example, in very rarefied gases, in which molecular collisions are infrequent; and in the boundary layers of a star, where radiation is passing energy to space; and for interacting fermions at very low temperature, where dissipative processes become ineffective.

When these 'cells' are defined, one admits that matter and energy may pass freely between contiguous 'cells', slowly enough to leave the 'cells' in their respective individual local thermodynamic equilibria with respect to intensive variables.

One can think here of two 'relaxation times' separated by order of magnitude. The longer relaxation time is of the order of magnitude of times taken for the macroscopic dynamical structure of the system to change. The shorter is of the order of magnitude of times taken for a single 'cell' to reach local thermodynamic equilibrium. If these two relaxation times are not well separated, then the classical non-equilibrium thermodynamical concept of local thermodynamic equilibrium loses its meaning. For example, in the atmosphere, the speed of sound is much greater than the wind speed; this favours the idea of local thermodynamic equilibrium of matter for atmospheric heat transfer studies at altitudes below about 60 km where sound propagates, but not above 100 km, where, because of the paucity of intermolecular collisions, sound does not propagate.

Milne's 1928 Definition of Local Thermodynamic Equilibrium in Terms of Radiative Equilibrium

Milne (1928), thinking about stars, gave a definition of 'local thermodynamic equilibrium' in terms of the thermal radiation of the matter in each small local 'cell'. He defined 'local thermodynamic equilibrium' in a 'cell' by requiring that it macroscopically absorb and spontaneously emit radiation as if it were in radiative equilibrium in a cavity at the temperature of the matter of the 'cell'. Then it strictly obeys Kirchhoff's law of equality of radiative emissivity and absorptivity, with a black body source function. The key to local thermodynamic equilibrium here is that the rate of collisions of ponderable matter particles such as molecules should far exceed the rates of creation and annihilation of photons.

Entropy in Evolving Systems

It is pointed out by W.T. Grandy Jr that entropy, though it may be defined for a non-equilibrium system, is when strictly considered, only a macroscopic quantity that refers to the whole system, and is not a dynamical variable and in general does not act as a local potential that describes local physical forces. Under special circumstances, however, one can metaphorically think as if the thermal variables behaved like local physical forces. The approximation that constitutes classical irreversible thermodynamics is built on this metaphoric thinking.

Speculated Thermodynamic Extremum Principles for Energy Dissipation and Entropy Production

Jou, Casas-Vazquez, Lebon (1993) note that classical non-equilibrium thermodynamics "has seen an extraordinary expansion since the second world war", and they refer to the Nobel prizes for work in the field awarded to Lars Onsager and Ilya Prigogine. Martyushev and Seleznev (2006) note the importance of entropy in the evolution of natural dynamical structures: "Great contribution has been done in this respect by two scientists, namely Clausius, ... , and Prigogine." Prigogine in his 1977 Nobel Lecture said: "... non-equilibrium may be a source of order. Irreversible processes may lead to a new type of dynamic states of matter which I have called "dissipative structures"." Glansdorff and Prigogine (1971) wrote on page xx: "Such 'symmetry breaking instabilities' are of special interest as they lead to a spontaneous 'self-organisation' of the system both from the point of view of its *space order* and its *function*."

Analysing the Rayleigh-Bénard convection cell phenomenon, Chandrashekhar (1961) wrote "Instability occurs at the minimum temperature gradient at which a balance can be maintained between the kinetic energy dissipated by viscosity and the internal energy released by the buoyancy force." With a temperature gradient greater than the minimum, viscosity can dissipate kinetic energy as fast as it is released by convection due to buoyancy, and a steady state with convection is stable. The steady state with convection is often a pattern of macroscopically visible hexagonal cells with convection up or down in the middle or at the 'walls' of each cell, depending on the temperature dependence of the quantities; in the atmosphere under various conditions it seems that either is possible. With a temperature gradient less than the minimum, viscosity and heat conduction are so effective that convection cannot keep going.

Glansdorff and Prigogine (1971) on page xv wrote "Dissipative structures have a quite different [from equilibrium structures] status: they are formed and maintained through the effect of exchange of energy and matter in non-equilibrium conditions." They were referring to the dissipation function of Rayleigh (1873) that was used also by Onsager (1931, I, 1931, II). On pages 78–80 of their book Glansdorff and Prigogine (1971) consider the stability of laminar flow that was pioneered by Helmholtz; they concluded that at a stable steady state of sufficiently slow laminar flow, the dissipation function was minimum. These advances have led to proposals for various extremal principles for the "self-organised" régimes that are possible for systems governed by classical linear and non-linear non-equilibrium thermodynamical laws, with stable stationary régimes being particularly investigated. Convection introduces effects of momentum which appear as non-linearity in the dynamical equations.

In the more restricted case of no convective motion, Prigogine wrote of "dissipative structures". Šilhavý (1997) offers the opinion that "... the extremum principles of [equilibrium] thermodynamics ... do not have any counterpart for [non-equilibrium] steady states (despite many claims in the literature)."

Prigogine's Proposed Theorem of Minimum Entropy Production

In 1945 Prigogine proposed a "Theorem of Minimum Entropy Production" which applies only to the linear regime near a stationary thermodynamically non-equilibrium state. The proof offered by Prigogine is open to serious criticism. A critical and unsupportive discussion of Prigogine's proposal is offered by Grandy (2008).

Speculated Principles of Maximum Entropy Production and Minimum Energy Dissipation

Onsager (1931, I) wrote: "Thus the vector field J of the heat flow is described by the condition that the rate of increase of entropy, less the dissipation function, be a maximum." Careful note needs to be taken of the opposite signs of the rate of entropy production and of the dissipation function, appearing in the left-hand side of Onsager's equation (5.13) on Onsager's page 423.

Although largely unnoticed at the time, Ziegler proposed an idea early with his work in the mechanics of plastics in 1961, and later in his book on thermomechanics revised in 1983, and in various papers (e.g., Ziegler (1987),). Ziegler never stated his principle as a universal law but he may have intuited this. He demonstrated his principle using vector space geometry based on an "orthogonality condition" which only worked in systems where the velocities were defined as a single vector or tensor, and thus, as he wrote at p. 347, was "impossible to test by means of macroscopic mechanical models", and was, as he pointed out, invalid in "compound systems where several elementary processes take place simultaneously".

In relation to the earth's atmospheric energy transport process, according to Tuck (2008), "On the macroscopic level, the way has been pioneered by a meteorologist (Paltridge 1975, 2001)." Initially Paltridge (1975) used the terminology "minimum entropy exchange", but after that, for example in Paltridge (1978), and in Paltridge (1979), he used the now current terminology "maximum entropy production" to describe the same thing. The logic of Paltridge's earlier work is open to serious criticism. Nicolis and Nicolis (1980) discuss Paltridge's work, and they comment that the behaviour of the entropy production is far from simple and universal. Later work by Paltridge focuses more on the idea of a dissipation function than on the idea of rate of production of entropy.

Sawada (1981), also in relation to the earth's atmospheric energy transport process, postulating a principle of largest amount of entropy increment per unit time, cites work in fluid mechanics by Malkus and Veronis (1958) as having "proven a principle of maximum heat current, which in turn is a maximum entropy production for a given boundary condition", but this inference is not logically valid. Again investigating planetary atmospheric dynamics, Shutts (1981) used an approach to the definition of entropy production, different from Paltridge's, to investigate a more abstract way to check the principle of maximum entropy production, and reported a good fit.

At present, for this area of investigation, the prospects for useful extremal principles seem clouded at best. C. Nicolis (1999) concludes that one model of atmospheric dynamics has an attractor which is not a regime of maximum or minimum dissipation; she says this seems to rule out the existence of a global organising principle, and comments that this is to some extent disappointing; she also points to the difficulty of finding a thermodynamically consistent form of entropy production. Another top expert offers an extensive discussion of the possibilities for principles of extrema of entropy production and of dissipation of energy: Grandy (2008) is very cautious, and finds difficulty in defining the 'rate of internal entropy production' in many cases, and finds that sometimes for the prediction of the course of a process, an extremum of the quantity called the rate of dissipation of energy may be more useful than that of the rate of entropy production; this quantity appeared in Onsager's 1931 origination of this subject.

Non-equilibrium thermodynamics has been successfully applied to describe biological systems such as Protein Folding/ unfolding and transport through membranes.

Competitive Inhibition

In competitive inhibition, the inhibitor and substrate compete for the enzyme (i.e., they can not bind at the same time). Often competitive inhibitors strongly resemble the real substrate of the enzyme. For example, methotrexate is a competitive inhibitor of the enzyme dihydrofolate reductase, which catalyses the reduction of dihydrofolate to tetrahydrofolate. In some cases, the inhibitor can bind to a site other than the binding-site of the usual substrate and exert an allosteric effect to change the shape of the usual binding-site. For example, strychnine acts as an allosteric inhibitor of the glycine receptor in the mammalian spinal cord and brain stem. Glycine is a major post-synaptic inhibitory neurotransmitter with a specific receptor site. Strychnine binds to an alternate site that reduces the affinity of the glycine receptor for glycine, resulting in convulsions due to lessened inhibition by the glycine. In competitive inhibition the maximal rate of the reaction is not changed, but higher substrate concentrations are required to reach a given maximum rate, increasing the apparent K_m.

Uncompetitive Inhibition

In uncompetitive inhibition, the inhibitor cannot bind to the free enzyme, only to the ES-complex. The EIS-complex thus formed is enzymatically inactive. This type of inhibition is rare, but may occur in multimeric enzymes.

Non-Competitive Inhibition

Non-competitive inhibitors can bind to the enzyme at the binding site at the same time as the substrate, but not to the active site. Both the EI and EIS complexes are enzymatically inactive. Because the inhibitor can not be driven from the enzyme by higher substrate concentration (in contrast to competitive inhibition), the apparent V_{max} changes. But because the substrate can still bind to the enzyme, the K_m stays the same.

Mixed Inhibition

This type of inhibition resembles the non-competitive, except that the EIS-complex has residual enzymatic activity. This type of inhibitor does not follow Michaelis-Menten equation.

In many organisms, inhibitors may act as part of a feedback mechanism. If an enzyme produces too much of one substance in the organism, that substance may act as an inhibitor for the enzyme at the beginning of the pathway that produces it, causing production of the substance to slow down or stop when there is sufficient amount. This is a form of negative feedback. Enzymes that are subject to this form of regulation are often multimeric and have allosteric binding sites for regulatory substances. Their substrate/velocity plots are not hyperbolar, but sigmoidal (S-shaped). Irreversible inhibitors react with the enzyme and form a covalent adduct with the protein. The inactivation is irreversible. These compounds include eflornithine a drug used to treat the parasitic disease sleeping sickness. Penicillin and Aspirin also act in this manner. With these drugs, the compound is bound in the active site and the enzyme then converts the inhibitor into an activated form that reacts irreversibly with one or more amino acid residues.

Uses of Inhibitors

Since inhibitors modulate the function of enzymes they are often used as drugs. A common example of an inhibitor that is used as a drug is aspirin, which inhibits the COX-1 and COX-2 enzymes that produce the inflammation messenger prostaglandin, thus suppressing pain and inflammation. However, other enzyme inhibitors are poisons. For example, the poison cyanide is an irreversible enzyme inhibitor that combines with the copper and iron in the active site of the enzyme cytochrome c oxidase and blocks cellular respiration.

Biological Function

Enzymes serve a wide variety of functions inside living organisms. They are indispensable for signal transduction and cell regulation, often via kinases and phosphatases. They also generate movement, with myosin hydrolyzing ATP to generate muscle contraction and also moving cargo around the cell as part of the cytoskeleton. Other ATPases in the cell membrane are ion pumps involved in active transport. Enzymes are also involved in more exotic functions, such as luciferase generating light in fireflies. Viruses can also contain enzymes for infecting cells, such as the HIV integrase and reverse transcriptase, or for viral release from cells, like the influenza virus neuraminidase. An important function of enzymes is in the digestive systems of animals. Enzymes such as amylases and proteases break down large molecules (starch or proteins, respectively) into smaller ones, so they can be absorbed by the intestines. Starch molecules, for example, are too large to be absorbed from the intestine, but enzymes hydrolyze the starch chains into smaller molecules such as maltose and eventually glucose, which can then be absorbed. Different enzymes digest different food substances. In ruminants, which have herbivorous diets, microorganisms in the gut produce another enzyme, cellulase, to break down the cellulose cell walls of plant fibre. Several enzymes can work together in a specific order, creating metabolic pathways. In a metabolic pathway, one enzyme takes the product of another enzyme as a substrate. After the catalytic reaction, the product is then passed on to another enzyme. Sometimes more than one enzyme can catalyse the same reaction in parallel; this can allow more complex regulation: with, for example, a low constant activity provided by one enzyme but an inducible high activity from a second enzyme.

Enzymes determine what steps occur in these pathways. Without enzymes, metabolism would neither progress through the same steps nor be fast enough to serve the needs of the cell. Indeed, a metabolic pathway such as glycolysis could not exist independently of enzymes. Glucose, for example, can react directly with ATP to become phosphorylated at one or more of its carbons. In the absence of enzymes, this occurs so slowly as to be insignificant. However, if hexokinase is added, these slow reactions continue to take place except that phosphorylation at carbon 6 occurs so rapidly that, if the mixture is tested a short time later, glucose-6-phosphate is found to be the only significant product. As a consequence, the network of metabolic pathways within each cell depends on the set of functional enzymes that are present.

Control of Activity

There are five main ways that enzyme activity is controlled in the cell.

1. Enzyme production (transcription and translation of enzyme genes) can be enhanced or diminished by a cell in response to changes in the cell's environment. This form of gene regulation is called enzyme induction and inhibition. For example, bacteria may become resistant to antibiotics such as penicillin because enzymes called beta-lactamases are induced that hydrolyze the crucial beta-lactam ring within the penicillin molecule. Another example are enzymes in the liver called cytochrome P450 oxidases, which are important in drug metabolism. Induction or inhibition of these enzymes can cause drug interactions.
2. Enzymes can be compartmentalized, with different metabolic pathways occurring in different cellular compartments. For example, fatty acids are synthesized

by one set of enzymes in the cytosol, endoplasmic reticulum and the Golgi apparatus and used by a different set of enzymes as a source of energy in the mitochondrion, through β-oxidation.

3. Enzymes can be regulated by inhibitors and activators. For example, the end product(s) of a metabolic pathway are often inhibitors for one of the first enzymes of the pathway (usually the first irreversible step, called *committed step*), thus regulating the amount of end product made by the pathways. Such a regulatory mechanism is called a negative feedback mechanism, because the amount of the end product produced is regulated by its own concentration. Negative feedback mechanism can effectively adjust the rate of synthesis of intermediate metabolites according to the demands of the cells. This helps allocate materials and energy economically, and prevents the manufacture of excess end products. The control of enzymatic action helps to maintain a stable internal environment in living organisms.

4. Enzymes can be regulated through post-translational modification. This can include phosphorylation, myristoylation and glycosylation. For example, in the response to insulin, the phosphorylation of multiple enzymes, including glycogen synthase, helps control the synthesis or degradation of glycogen and allows the cell to respond to changes in blood sugar. Another example of post-translational modification is the cleavage of the polypeptide chain. Chymotrypsin, a digestive protease, is produced in inactive form as chymotrypsinogen in the pancreas and transported in this form to the stomach where it is activated. This stops the enzyme from digesting the pancreas or other tissues before it enters the gut. This type of inactive precursor to an enzyme is known as a zymogen.

5. Some enzymes may become activated when localized to a different environment (e.g., from a reducing (cytoplasm) to an oxidizing (periplasm) environment, high pH to low pH, etc.). For example, hemagglutinin in the influenza virus is activated by a conformational

change caused by the acidic conditions, these occur when it is taken up inside its host cell and enters the lysosome.

Involvement in Disease

Since the tight control of enzyme activity is essential for homeostasis, any malfunction (mutation, overproduction, underproduction or deletion) of a single critical enzyme can lead to a genetic disease. The importance of enzymes is shown by the fact that a lethal illness can be caused by the malfunction of just one type of enzyme out of the thousands of types present in our bodies. One example is the most common type of phenylketonuria. A mutation of a single amino acid in the enzyme phenylalanine hydroxylase, which catalyses the first step in the degradation of phenylalanine, results in build-up of phenylalanine and related products. This can lead to mental retardation if the disease is untreated. Another example of enzyme deficiency is pseudocholinesterase, in which there is slow metabolic degradation of exogenous choline. Another example is when germline mutations in genes coding for DNA repair enzymes cause hereditary cancer syndromes such as xeroderma pigmentosum. Defects in these enzymes cause cancer since the body is less able to repair mutations in the genome. This causes a slow accumulation of mutations and results in the development of many types of cancer in the sufferer. Oral administration of enzymes can be used to treat several diseases (e.g. pancreatic insufficiency and lactose intolerance). Since enzymes are proteins themselves they are potentially subject to inactivation and digestion in the gastrointestinal environment. Therefore a non-invasive imaging assay was developed to monitor gastrointestinal activity of exogenous enzymes (prolyl endopeptidase as potential adjuvant therapy for celiac disease) in vivo.

Naming Conventions

An enzyme's name is often derived from its substrate or the chemical reaction it catalyses, with the word ending in *-ase*. Examples are lactase, alcohol dehydrogenase and DNA polymerase. This may result in different enzymes, called isozymes, with the same function having the same basic name. Isoenzymes have a different amino acid sequence and might be distinguished by their optimal pH, kinetic properties or immunologically. Isoenzyme and isozyme are homologous proteins. Furthermore, the normal physiological reaction an enzyme catalyses may not be the same as under artificial conditions.

This can result in the same enzyme being identified with two different names. For example, glucose isomerase, which is used industrially to convert glucose into the sweetener fructose, is a xylose isomerase *in vivo* (within the body).

The International Union of Biochemistry and Molecular Biology have developed a nomenclature for enzymes, the EC numbers; each enzyme is described by a sequence of four numbers preceded by "EC". The first number broadly classifies the enzyme based on its mechanism.

The top-level classification is

- EC 1 *Oxidoreductases*: catalyse oxidation/reduction reactions
- EC 2 *Transferases*: transfer a functional group (*e.g.* a methyl or phosphate group)
- EC 3 *Hydrolases*: catalyse the hydrolysis of various bonds
- EC 4 *Lyases*: cleave various bonds by means other than hydrolysis and oxidation
- EC 5 *Isomerases*: catalyse isomerization changes within a single molecule
- EC 6 *Ligases*: join two molecules with covalent bonds.

According to the naming conventions, enzymes are generally classified into six main family classes and many sub-family classes. Some web-servers, e.g., EzyPred and bioinformatics tools have been developed to predict which main family class and sub-family class an enzyme molecule belongs to according to its sequence information alone via the pseudo amino acid composition.

Industrial Applications

Enzymes are used in the chemical industry and other industrial applications when extremely specific catalysts are required. However, enzymes in general are limited in the number of reactions they have evolved to catalyse and also by their lack of stability in organic solvents and at high temperatures. As a consequence, protein engineering is an active area of research and involves attempts to create new enzymes with novel properties, either through rational design or *in vitro* evolution. These efforts have begun to be successful, and a few enzymes have now been designed "from scratch" to catalyse reactions that do not occur in nature.

Hemp

Hemp (from Old English *hænep*) is a term reserved mainly for low tetrahydr-ocannabinol (THC) strains of the plant *Cannabis sativa*. Of the approximately 2000 cannabis plants varieties known, about 90% contain only low-grade THC and are most useful for their fibre, seeds and medicinal or psychoactive oils. Hemp is one of the earliest domesticated plants known.

In modern times hemp is used for industrial purposes including paper, textiles, clothing, biodegradable plastics, construction (as with Hemcrete and insulation), body products, health food and bio-fuel. Hemp is thus legally grown in many countries across the world including Spain, China, Japan, Korea, France, North Africa and Ireland. Although hemp is commonly associated with marijuana (hemp's THC-rich cousin), since 2007 the commercial success of hemp food products has grown considerably. Hemp is one of the faster growing biomasses known, producing up to 25 tonnes of dry matter per hectare per year. A typical average yield in large scale modern agriculture is about 2.5–3.5/ t/ac (air dry stem yields of dry, retted stalks per acre at 12% moisture). Approximately one tonne of bast fibre and 2–3 tonnes of core material can be decorticated from 3–4 tonnes of good quality, dry retted straw.

Hemp is very environmentally friendly as it requires few pesticides and no herbicides. It has been called a carbon-negative raw material. Results indicate that high yield of hemp may require high total nutrient levels (field plus fertilizer nutrients) similar to a high yielding wheat crop. *Cannabis sativa* L. subsp. *sativa* var. *sativa* is the variety grown for industrial use, while *C. sativa* subsp. *indica* generally has poor fibre quality and is primarily used for recreational and medicinal purposes. The major difference between the two types of plants is the appearance and the amount of Δ^9-tetrahydrocannabinol (THC) secreted in a resinous mixture by epidermal hairs called glandular trichomes, although they can also be distinguished genetically. Oilseed and fibre varieties of *Cannabis* approved for industrial hemp production produce only minute amounts of this psychoactive drug, not enough for any physical or psychological effects. Typically, hemp contains below 0.3% THC, while cultivars of *Cannabis* grown for recreational use can contain anywhere from 2% to over 20%.

The world leading producer of hemp is China with smaller production in Europe, Chile and North Korea. While more hemp is exported to the United States than to any other country, the United States Government does not consistently distinguish between marijuana and the non-psychoactive *Cannabis* used for industrial and commercial purposes.

Hemp is used for a wide variety of purposes including the manufacture of cordage of varying tensile strength, durable clothing and nutritional products. The *bast fibres* can be used in 100% hemp products, but are commonly blended with other organic fibres such as flax, cotton or silk, for apparel and furnishings, most commonly at a 55%/45% hemp/cotton blend. The inner two fibres of hemp are more woody and are more often used in non-woven items and other industrial applications, such as mulch, animal bedding and litter.

The oil from the fruits ("seeds") oxidizes (commonly, though inaccurately, called "drying") to become solid on exposure to air, similar to linseed oil, and is sometimes used in the manufacture of oil-based paints, in creams as a moisturizing agent, for cooking, and in plastics. Hemp seeds have been used in bird seed mix as well. A survey in 2003 showed that more than 95% of hemp seed sold in the EU was used for animal feed (bird seed, bait for fishing). Hempseed is also used as a fishing bait.

Food

Hemp seeds can be eaten raw, ground into a meal, sprouted, made into hemp milk (akin to soy milk), prepared as tea, and used in baking. The fresh leaves can also be consumed in salads. Products include cereals, frozen waffles, hemp tofu, and nut butters. A few companies produce value added hemp seed items that include the seed oils, whole hemp grain (which is sterilized by law in the United States, where they import it from China and Canada), dehulled hemp seed (the whole seed without the mineral rich outer shell), hemp flour, hemp cake (a by-product of pressing the seed for oil) and hemp protein powder. Hemp is also used in some organic cereals, for non-dairy milk somewhat similar to soy and nut milks, and for non-dairy hemp "ice cream." Within the UK, the Department for Environment, Food and Rural Affairs (Defra) has treated hemp as purely a non-food crop. Seed appears on the UK market as a legal food product, and cultivation licenses are available for this purpose. In North America, hemp seed food products are sold, typically in health food stores or through mail order. The United States Department of Agriculture estimates that "the market potential for hemp seed as a food ingredient is unknown. However, it probably will remain a small market, like those for sesame and poppy seeds."

The first identified coarse paper, made from hemp, dates to the early Western Han Dynasty, two hundred years before the nominal invention of papermaking by Cai Lun, who improved and standardized paper production using a range of inexpensive materials, including hemp ends, approximately 2000 years ago. Recycled hemp clothing, rags and fishing nets were used as inputs for paper production. The Saint Petersburg, Russia paper mill of Goznak opened in 1818. It used hemp as its main input material. Paper from the mill was used in the printing of "bank notes, stamped paper, credit bills, postal stamps, bonds, stocks, and other watermarked paper." In 1916, U.S. Department of Agriculture chief scientists Lyster Hoxie Dewey and Jason L. Merrill created paper made from hemp pulp and concluded that paper from hemp hurds was "favourable in comparison with those used with pulp wood." Modern research has not confirmed the positive finding about hemp hurds. They are only 32% and 38% cellulose. The actual production of hemp fibre in the U.S continued to decline until 1933 to around 500 tons/year. Between 1934-35, the cultivation of hemp began to increase but still at a very low level and with no significant increase of paper from hemp.

Hemp has never been used for commercial high-volume paper production due to its relatively high processing cost. Currently there is a small niche market for hemp pulp, for example as cigarette paper. Hemp fibre is mixed with fibre from other sources than hemp. In 1994 there was no significant production of 100% true hemp paper. World hemp pulp production was believed to be around 120,000 tons per year in 1991 which was about 0.05% of the world's annual pulp production volume. The total world production of hemp fibre had in 2003 declined to about 60 000 to 80 000 tons. This can be compared to a typical pulp mill for wood fibre, which is never smaller than 250,000 tons per annum. The cost of hemp pulp is approximately six times that of wood pulp, mostly because of the small size and outdated equipment of the few hemp processing plants in the Western world, and because hemp is harvested once a year (during August) and needs to be stored to feed the mill the whole year through. This storage requires a lot of (mostly manual) handling of the bulky stalk bundles. Another issue is that the entire hemp plant cannot be economically prepared for paper production. While the wood products industry uses nearly 100% of the fibre from harvested trees, only about 25% of the dried hemp stem — the bark, called bast — contains the long, strong fibres desirable for paper production. All this accounts for a high raw material cost. Hemp pulp is bleached with hydrogen peroxide, a process today also commonly used for wood pulp.

Today, hemp paper for cigarettes, among other things, is produced in Spain and the UK. Around the year 2000, the production quantity of flax and hemp pulp total 25000-30000 tons per year, having been produced from approximately 37000-45000 tonnes fibres. Up to 80% of the produced pulp is used for speciality papers (including 95% of cigarette paper). Only about 20% hemp fibre input goes into the standard pulp area and are here mostly in lower quality (untreated oakum high shive content added) wood pulps. With hemp pulp alone, the proportion of speciality papers probably at about 99%. The market is considered saturated with little or no growth in this area.

Hemp jewellery is the product of knotting hemp twine through the practice of macramé. Hemp jewellery includes bracelets, necklaces, anklets, rings, watches and other adornments. Some jewellery features beads made from glass, stone, wood and bones. The hemp twine varies in thickness and comes in a variety of colours. There are many different stitches used to create hemp jewellery, however, the half knot and full knot stitches are most common. Hemp rope was used in the age of sailing ships, though the rope had to be protected by tarring, since hemp rope has a propensity for breaking from rot, as the capillary effect of the rope-woven fibres tended to hold liquid at the interior, while seeming dry from the outside. Tarring was a labour-intensive process, and earned sailors the nickname "Jack Tar". Hemp rope was phased out when Manila, which does not require tarring, became widely available. Manila is sometimes referred to as Manila hemp, but is not related to hemp; it is abaca, a species of banana.

Hemp shives are the core of the stem, hemp hurds are broken parts of the core. In the EU, they are used for animal bedding (horses, for instance), or for horticultural mulch. Industrial hemp is much more profitable if both fibres and shives (or even seeds) can be used.

Hemp can be used as a "mop crop" to clear impurities out of wastewater, such as sewage effluent, excessive phosphorus from chicken litter, or other unwanted substances or chemicals. Eco-technologist Dr. Keith Bolton from Southern Cross University in Lismore, New South Wales, Australia, is a leading researcher in this area. Hemp is being used to clean contaminants at the Chernobyl nuclear disaster site. This is known as phytoremediation - the process by the cleaning radiation as well as a variety of other toxins from the soil, water, and air. Hemp, because of its height, dense foliage and its high planting density as a crop, is a very effective and long used method of killing tough weeds in farming by minimising the pool of weed seeds of the soil. Using hemp this way can help farmers avoid the use of herbicides, to help gain organic certification and to gain the benefits of crop rotation *per se*. Due to its rapid, dense growth characteristics, in some jurisdictions hemp is considered a prohibited noxious weed, much like Scotch Broom. It has been used extensively to kill weeds in agriculture. Biofuels, such as biodiesel and alcohol fuel, can be made from the oils in hemp seeds and stalks, and the fermentation of the plant as a whole, respectively. Biodiesel produced from hemp is sometimes known as "hempoline". Filtered hemp oil can be used directly to power diesel engines. In 1892, Rudolf Diesel invented the diesel engine, which he intended to fuel "by a variety of fuels, especially vegetable and seed oils, which earlier were used for oil lamps, i.e. the Argand lamp." Production of vehicle fuel from hemp is very small. Commercial biodiesel and biogas is typically produced from cereals, coconuts, palmseeds and cheaper raw materials like garbage, wastewater, dead plant and animal material, animal feces and kitchen waste. Hemp is usually planted between March and May in the northern hemisphere, between September and November in the southern hemisphere. It matures in about three to four months. Millennia of selective breeding have resulted in varieties that look quite different. Also, breeding since circa 1930 has focused quite specifically on producing strains which would perform very poorly as sources of drug material. Hemp grown for fibre is planted closely, resulting in tall, slender plants with long fibres. Ideally, according to Britain's Department for Environment, Food and Rural Affairs, the herb should be harvested before it flowers. This early cropping is done because fibre quality declines if flowering is allowed and, incidentally, this cropping also pre-empts the herb's maturity as a potential source of drug material. However, in these strains of industrial hemp the tetrahydrocannabinol (THC) content would have been very low, regardless.

The seeds are sown from mid April to mid May with grain drills to 4–6 cm sowing depth. Hemp needs less fertilizer than corn does. A total of 60–150 kg of nitrogen, 40–140 kg phosphorus (P_2O_5) and 75–200 kg of potassium [5] per acre for hemp fibre made before sowing and again later, maybe three to four weeks . When practiced, especially in France double use of fibre and seed fertilization with nitrogen doses up to 100 kg / ha rather low. Organic fertilizers such as manure can utilise industrial hemp well. Neither weeds nor crop protection measures are necessary.

Smallholder plots are usually harvested by hand. The plants are cut at 2 to 3 cm above the soil and left on the ground to dry. Mechanical harvesting is now common, using specially adapted cutter-binders or simpler cutters. The cut hemp is laid in swathes to dry for up to four days. This was traditionally followed by *retting*, either water retting (the bundled hemp floats in water) or dew retting (the hemp remains on the ground and is affected by the moisture in dew, and by molds and bacterial action). Modern processes use steam and machinery to separate the fibre, a process known as thermomechanical pulping.

For profitable hemp farming, particularly deep, humus-rich, nutrient-rich soil with controlled water flow is preferable. Water logged acidic, compressed or extremely light (sandy) soils primarily affect the early development of plants. Steep slopes and high altitudes of more than 400 m above sea level are best avoided. Hemp is relatively insensitive to cold temperatures and can withstand frost down to -5 degrees C. Seeds can germinate down to 1-3 degrees. Hemp needs a lot of heat, so earlier varieties come to maturation. The water requirement is 300-500 l / kg dry matter is relatively high, up to 3 feet growing roots into the soil can also use water supplies from deeper soil layers.

Hemp benefits crops grown after it. For this reason it is generally grown before winter cereals. Advantageous changes are high weed suppression, soil loosening by the large hemp root system and the positive effect on soil tilth. Since hemp is very self-compatible, it can also be grown several years in a row in the same fields (monoculture). Hemp has been grown for millennia in Asia and the Middle East for its fibre. Commercial production of hemp in the West took off in the eighteenth century, but was grown in the sixteenth century in eastern England. Because of colonial and naval expansion of the era, economies needed large quantities of hemp for rope and oakum. Other important producing countries were China, North Korea, Hungary, the former Yugoslavia, Romania, Poland, France and Italy. In Western Europe, nobody banned the cultivation of hemp in the 1930s but the commercial cultivation ceased almost anyhow in the decades after the 1930s. Hemp was simply ousted by artificial fibres. From the 1950s to the 1980s, the Soviet Union was the world's largest producer (3,000 km^2 in 1970). The main production areas were in Ukraine, the Kursk and Orel regions of Russia, and near the Polish border. Since its inception in 1931, the Hemp Breeding Department at the Institute of Bast Crops in Hlukhiv (Glukhov), Ukraine, has been one of the world's largest centres for developing new hemp varieties, focusing on improving fibre quality, per-hectare yields, and low THC content. In Japan, hemp was historically used as paper and a fibre crop. There is archaeological evidence cannabis was used for clothing and the seeds were eaten in Japan back to the Jômon period (10,000 to 300 BCE). Many Kimono designs portray hemp, or *asa* (Japanese: »•), as a beautiful plant. In 1948, marijuana was restricted as a narcotic drug. The ban on marijuana imposed by the United States authorities was alien to Japanese culture, as the drug had never been widely used in Japan before. Though these laws against marijuana are some of the world's strictest, allowing five years imprisonment for possession of the drug, they exempt hemp growers, whose crop is used to make robes for Buddhist monks and loincloths for Sumo wrestlers. Because marijuana use in Japan has doubled in the past decade, these exemptions have recently been called into question.

The cultivation of hemp in Portuguese lands began around the fourteenth century onwards, it was raw material for the preparation of rope and plugs for the Portuguese ships. Colonies for factories for the production of flax hemp, such as the Royal Flax Hemp Factory in Brazil. After the Restoration of Independence in 1640, in order to recover the ailing Portuguese naval fleet, were encouraged its cultivation as the Royal Decree of D. John IV in 1656 . At that time its cultivation was carried out in Tras-os-Montes, Zone Tower Moncorvo, more precisely in Vilariça Valley, fertile land for any crop irrigation, and a very large area, flat and very fertile culture still wide until the last century grew up tobacco, a plant that needs a large space to expand and grow, the area lies in the valley of Serra de Bornes.

In 1971 this cultivar is considered illegal because of marijuana, a decision subsequently revoked by the European Union . Air dry stem yields in Ontario have from 1998 and onward ranged from 2.6-14.0 tonnes of dry, retted stalks per hectare (1-5.5 t/ac) at 12% moisture. Yields in Kent County, have averaged 8.75 t/ha (3.5 t/ac). Northern Ontario crops averaged 6.1 t/ha (2.5 t/ac) in 1998. Statistic for the European Union for 2008 to 2010 say that the average yield of hemp straw has varied between 6.3 and 7.3 ton per ha. Only a part of that is bast fibre. Approximately one tonne of bast fibre and 2-3 tonnes of core material can be decorticated from 3-4 tonnes of good quality, dry retted straw. For an annual yield of this level is it in Ontario recommended to add nitrogen (N):70–110 kg/ha, phosphate (P2O5): up to 80 kg/ha and potash (K2O): 40–90 kg/ha. The average yield of dry hemp stalks in Europe was 6 ton/ha (2.4 ton/ac) in 2001 and 2002.

FAO argue that an optimum yield of hemp fibre is more than 2 tonnes per ha, while average yields are around 650 kg/ha. There are a lot of very uncertain, easily misleading and sometimes clearly incorrect numbers about the yield from hemp in ton/hectare or pounds/acre etc. on the Internet. Frequently it is not specified if the numbers describe: the total biomass; the total yield of biomass; the total yield of dried stalk; or the total yield of fibre from the bark. Furthermore, it is frequently not specified whether the numbers measure wet or dry material. Hemp can contain a lot of water. In modern industrial agriculture is about 42% of the plants' biomass returned to the soil in the form of leaves, roots and tops. A total of 41 varieties of hemp with low levels of tetrahydrocannabinol (THC) are certified by the European Union (EU). They have, unlike other types, a very high fibre content of 30-40%. In contrast to cannabis for medical use, varieties grown for fibre and seed have less than 0.2% THC and they are unsuitable for producing hashish and marijuana. The most important cannabinoid in industrial hemp is cannabidiol (CBD) with a proportion of 1 to 5%.

There are broadly three groups of *Cannabis* varieties being cultivated today:

- Varieties primarily cultivated for their fibre, characterised by long stems and little branching, extreme red, yellow, blue or purple colouration, or thickness of stem and solid core, such as hemp *Cannabis oglalas*, and more generally called industrial hemp.
- Varieties grown for hemp seed oil which is high in protein and essential fatty acids and has no psychoactive properties.
- Varieties grown for medicinal, spiritual or recreational purposes.

A nominal, if not legal distinction is often made between hemp, with concentrations of the psychoactive chemical THC far too low to be useful as a drug, and *Cannabis* used for medical, recreational, or spiritual purposes. Hemp plants can be vulnerable to various pathogens, including bacteria, fungi, nematodes, viruses and other miscellaneous pathogens. Such diseases often lead to reduced fibre quality, stunted growth, and death of the plant. hese diseases rarely affect the yield of a hemp field, so hemp production is not traditionally dependent on the use of pesticides. Hemp use archaeologically dates back to the Neolithic Age in China, with hemp fibre imprints found on Yangshao culture pottery dating from the 5th century BC. The Chinese later used hemp to make clothes, shoes, ropes, and an early form of paper.

The classical Greek historian Herodotus (ca. 480 BC) reported that the inhabitants of Scythia would often inhale the vapors of hemp-seed smoke, both as ritual and for their own pleasurable recreation.

Textile expert Elizabeth Wayland Barber summarises the historical evidence that *Cannabis sativa*, "grew and was known in the Neolithic period all across the northern latitudes, from Europe (Germany, Switzerland, Austria, Romania, the Ukraine) to East Asia (Tibet and China)," but, "textile use of Cannabis sativa does not surface for certain in the West until relatively late, namely the Iron Age." "I strongly suspect, however, that what catapulted hemp to sudden fame and fortune as a cultigen and caused it to spread rapidly westwards in the first millennium B.C. was the spread of the habit of pot-smoking from somewhere in south-central Asia, where the drug-bearing variety of the plant originally occurred. The linguistic evidence strongly supports this theory, both as to time and direction of spread and as to cause."

Jews living in Palestine in the 2nd century were familiar with the cultivation of hemp, as witnessed by a reference to it in the Mishna (*Kil'ayim* 2:5) as a variety of plant, along with Arum, that sometimes takes as many as three years to grow from a seedling.

In late medieval Germany and Italy, hemp was employed in cooked dishes, as filling in pies and tortes, or boiled in a soup.

Hemp in later Europe was mainly cultivated for its fibres, and was used for ropes on many ships, including those of Christopher Columbus. The use of hemp as a cloth was centred largely in the countryside, with higher quality textiles being available in the towns.

The Spaniards brought hemp to the Western Hemisphere and cultivated it in Chile starting about 1545. However, in May 1607, "hempe" was among the crops Gabriel Archer observed being cultivated by the natives at the main Powhatan village, where Richmond, Virginia is now situated; and in 1613, Samuell Argall reported wild hemp "better than that in England" growing along the shores of the upper Potomac. As early as 1619, the first Virginia House of Burgesses passed an Act requiring all planters in Virginia to sow "both English and Indian" hemp on their plantations. The Puritans are first known to have cultivated hemp in New England in 1645. In 1937, the Marihuana Tax Act of 1937 was passed in the United States. It levied a tax on anyone who dealt commercially in cannabis, hemp, or marijuana. It was repealed by an overriding law in 1970.

Hemp was used extensively by the United States during World War II. Uniforms, canvas, and rope were among the main textiles created from the hemp plant at this time. Much of the hemp used was cultivated in Kentucky and the Midwest.

Historically, hemp production had made up a significant portion of antebellum Kentucky's economy. Before the American Civil War, many slaves worked on plantations producing hemp.

During World War II, the U.S. produced a short 1942 film, *Hemp for Victory*, promoting hemp as a necessary crop to win the war.

By the early twentieth century, the advent of the steam engine and the Diesel engine ended the reign of the sailing ship. The production of iron and steel for cable and ships' hulls further eliminated natural fibres in marine use. Hemp had long since fallen out of favour in the sailing industry in preference to Manila hemp.

The world-leading producer of hemp is China, with smaller production in Europe, Chile and North Korea. Over thirty countries produce industrial hemp, including Australia, Austria, Canada, Chile, China, Denmark, Egypt, Finland, France, Germany, Great Britain, Hungary, India, Italy, Japan, Korea, Netherlands, New Zealand, Poland, Portugal, Romania, Russia, Slovenia, Spain, Sweden, Switzerland, Thailand, Turkey and

Ukraine. The United Kingdom, and Germany all resumed commercial production in the 1990s. British production is mostly used as bedding for horses; other uses are under development. Companies in Canada, the UK, the United States and Germany, among many others, process hemp seed into a growing range of food products and cosmetics; many traditional growing countries still continue to produce textile-grade fibre.

In the Australian states of Victoria, Queensland and, most recently, New South Wales the state governments have issued licences to grow hemp for industrial use. The state of Victoria was an early adopter in 1998, and has reissued the regulation in 2008. Queensland has allowed industrial production under licence since 2002, where the issuance is controlled under the Drugs Misuse Act 1986. Most recently, New South Wales now issues licences under a law, the Hemp Industry Regulations Act 2008 (No 58), that came into effect as of 6 November 2008.

Commercial production (including cultivation) of industrial hemp has been permitted in Canada since 1998 under licenses and authorization issued by Health Canada (9,725 ha in 2004, 5450 ha in 2009). France is Europe's biggest producer with 8,000 hectares cultivated. 70-80 % of the hemp fibre produced in Europe in 2003 was used for speciality pulp for cigarette papers and technical applications. About 15% is used in the automotive sector and 5-6% were used for insulation mats. Approximately 95% of hurds were used as animal bedding, while almost 5% were used in the building sector. In 2010/2011 was the total area cultivated with hemp in the EU about 11 000 ha, a decline compared with previous year. Licences for hemp cultivation are issued in the European Union, Canada, in all states of Australia and ten states in the United States. In the United Kingdom, these licences are issued by the Home Office under the Misuse of Drugs Act 1971. When grown for non-drug purposes, hemp is referred to as industrial hemp, and a common product is fibre for use in a wide variety of products, as well as the seed for nutritional aspects and for the oil. Feral hemp or ditch weed is usually a naturalized fibre or oilseed strain of *Cannabis* that has escaped from cultivation and is self-seeding. Vermont and North Dakota have passed laws enabling hemp licensure. Both states are waiting for permission to grow hemp from the DEA. Currently, North Dakota representatives are pursuing legal measures to force DEA approval. Oregon has licensed industrial hemp as of August 2009. Hemp is not legal to grow in the U.S. under Federal law because of its relation to marijuana, and any imported hemp products must meet a zero tolerance level. It is considered a controlled substance under the Controlled Substances Act (P.L. 91-513; 21 U.S.C. 801 et seq.). Some states have made the cultivation of industrial hemp legalÿ but these states — North Dakota, Hawaii, Kentucky, Maine, Maryland, Oregon, California, Montana, West Virginia and Vermont — have not yet begun to grow it because of resistance from the federal Drug Enforcement Administration.

Uruguay has also approved a project of hemp production as of the second half of 2010.

Chapter 9

Nomenclature

The IUPAC nomenclature of organic chemistry is a systematic method of naming organic chemical compounds as recommended by the International Union of Pure and Applied Chemistry (IUPAC). Ideally, every possible organic compound should have a name from which an unambiguous structural formula can be drawn. There is also an IUPAC nomenclature of inorganic chemistry. For ordinary communication, to spare a tedious description, the official IUPAC naming recommendations are not always followed in practice except when it is necessary to give a concise definition to a compound, or when the IUPAC name is simpler (e.g. ethanol against ethyl alcohol). Otherwise the common or trivial name may be used, often derived from the source of the compound. In addition, very long names may actually be less concise than structural formulae.

Basic Principles

In'chemistry, a number of prefixes, suffixes and infixes are used to describe the type and position of functional groups in the compound. The steps to naming an organic compound are:

1. Identify the parent hydrocarbon chain. This chain must follow the following rules, in order of precedence:
    1. It should have maximum substituents of the suffix functional group. By suffix, it is meant that the parent functional group should have a suffix, unlike halogen substituents. If more than one functional group is present, use the one with highest precedence as shown here.

   2. It should have maximum number of multiple bonds
   3. It should have maximum number of double bonds.
   4. It should have the maximum length.
2. Identify the parent functional group, if any, with the highest order of precedence.
3. Identify the side-chains. *Side chains are the carbon chains that are not in the parent chain, but are branched off from it.*
4. Identify the remaining functional groups, if any, and name them by the name of their ions (such as hydroxy for -OH, oxy for =O, oxyalkane for O-R, etc.). Different side-chains and functional groups will be grouped together in alphabetical order. (The prefixes di-, tri-, etc. are not taken into consideration for grouping alphabetically. For example, ethyl comes before dihydroxy or dimethyl, as the "e" in "ethyl" precedes the "h" in "dihydroxy" and the "m" in "dimethyl" alphabetically. The "di" is not considered in either case). In the case of there being both side chains and secondary functional groups, they should be written mixed together in one group rather than in two separate groups.
5. Identify double/triple bonds.
6. Number the chain. To number the chain, first number in both directions (left to right and right to left), and then choose the numbering which follows these rules, in order of precedence:
   1. Has the lowest locant (or locants) for the suffix functional group. Locants are the numbers on the carbons to which the substituent is directly attached.
   2. Has the lowest locants for multiple bonds (The locant of a multiple bond is the number of the adjacent carbon with a lower number).
   3. Has the lowest locants for double bonds
   4. Has the lowest locants for prefixes.
7. Number the various substituents and bonds with their locants. If there is more than one of the same type of substituent/double bond, add the prefix (di-, tri-, etc.) before it. The numbers for that type of side chain will

be grouped in ascending order and written before the name of the side-chain. If there are two side-chains with the same alpha carbon, the number will be written twice. Example: 2,2,3-trimethyl- . If there are both double bonds and triple bonds, write the "ene" before the "yne". In case the main functional group is a terminal functional group (A group which can only exist at the end of a chain, like formyl and carboxyl groups), there is no need to number it.

8. Arrange everything like this: Group of side chains and secondary functional groups with numbers made in step 3 + prefix of parent hydrocarbon chain (eth, meth) + double/ triple bonds with numbers (or "ane") + primary functional group suffix with numbers. Wherever it says "with numbers", it is understood that between the word and the numbers, you use the prefix(di-, tri-)
9. Add punctuation:
    1. Put commas between numbers (2 5 5 becomes 2,5,5)
    2. Put a hyphen between a number and a letter (2 5 5 trimethylheptane becomes 2,5,5-trimethylheptane)
    3. Successive words are merged into one word (trimethyl heptane becomes trimethylheptane)

Now, we go by the steps:

1. The parent hydrocarbon chain has 23 carbons. It is called tricos-.
2. The functional groups with the highest precedence are the two ketone groups.
    1. The groups are on carbon atoms 3 & 9. As there are two, we write 3,9-dione.
    2. The numbering of the molecule is based on the ketone groups. When numbering from left to right, the ketone groups get numbered 3 and 9. When numbering from right to left, the ketone groups get numbered 15 and 21. 3 is less than 15, therefore, the numbering is done left to right, and the ketones are numbered 3 & 9. The lesser number is always used, not the sum of the constituents numbers.

3. The side chains are: an ethyl- at carbon 4, an ethyl- at carbon 8, and a butyl- at carbon 12.

Note: The -O-CH_3 at carbon atom 15 is not a side chain, but it is a methoxy functional group

- o There are two ethyl- groups, so they are combined to create, 4,8-diethyl.
- o The side chains shall be grouped like this: 12-butyl-4,8-diethyl. (But this is not the final grouping, as functional groups may be added in between.)

4. The secondary functional groups are: a hydroxy- at carbon 5, a chloro- at carbon 11, a methoxy- at carbon 15, and a bromo- at carbon 18. Grouped with the side chains, we get 18-bromo-12-butyl-11-chloro-4,8-diethyl-5-hydroxy-15-methoxy
5. There are two double bonds: one between carbons 6 & 7, and one between carbons 13 & 14. They will be called 6,13-diene. There is one triple bond between carbon atoms 19 & 20. It will be called 19-yne
6. The arrangement (with punctuation) is: 18-bromo-12-butyl-11-chloro-4,8-diethyl-5-hydroxy-15-methoxytricos-6,13-diene-19-yne-3,9-dione

The final name is 18-bromo-12-butyl-11-chloro-4,8-diethyl-5-hydroxy-15-methoxytricos-6,13-diene-19-yne-3,9-dione.

Nomenclature

The IUPAC nomenclature (systematic way of naming compounds) for alkanes is based on identifying hydrocarbon chains. Unbranched, saturated hydrocarbon chains are named systematically with a Greek numerical prefix denoting the number of carbons and the suffix "-ane". August Wilhelm von Hofmann suggested systematizing nomenclature by using the whole sequence of vowels a, e, i, o and u to create suffixes -ane, -ene, -ine (or -yne), -one, -une, for the hydrocarbons. The first three name hydrocarbons with single, double and triple bonds; "-one" represents a ketone; "-ol" represents an alcohol or OH group; "-oxy-" means an ether and refers to oxygen between two carbons, so that methoxy-methane is the IUPAC name for dimethyl ether. It is difficult or impossible to find compounds with more than one IUPAC name. This is because shorter chains attached to longer chains are prefixes and the convention includes brackets. Numbers in the name, referring to which carbon a group is attached to, should be as low as possible, so that 1- is implied and usually omitted from names of organic compounds with only one side-group. Symmetric compounds will have two ways of arriving at the same name.

Linear Alkanes

Straight-chain alkanes are sometimes indicated by the prefix *n-* (for *normal*) where a non-linear isomer exists. Although this is not strictly necessary, the usage is still common in cases where there is an important difference in properties between the straight-chain and branched-chain isomers, e.g., *n*-hexane or 2- or 3-methylpentane.

The members of the series (in terms of number of carbon atoms) are named as follows:

methane, CH_4 - one carbon and four hydrogen

ethane, C_2H_6 - two carbon and six hydrogen

propane, C_3H_8 - three carbon and 8 hydrogen

butane, C_4H_{10} - four carbon and 10 hydrogen

pentane, C_5H_{12} - five carbon and 12 hydrogen

hexane, C_6H_{14} - six carbon and 14 hydrogen

These names were derived from methanol, ether, propionic acid and butyric acid, respectively. Alkanes with five or more carbon atoms are named by adding the suffix -ane to the appropriate numerical multiplier prefix with elision of any terminal vowel (*-a* or *-o*) from the basic numerical term. Hence, pentane, C_5H_{12}; hexane, C_6H_{14}; heptane, C_7H_{16}; octane, C_8H_{18}; etc. The prefix is generally Greek, with the exceptions of nonane which has a Latin prefix, and undecane and tridecane which have mixed-language prefixes.

Branched Alkanes

Simple branched alkanes often have a common name using a prefix to distinguish them from linear alkanes, for example *n*-pentane, isopentane, and neopentane.

IUPAC naming conventions can be used to produce a systematic name.

The key steps in the naming of more complicated branched alkanes are as follows:

- Identify the longest continuous chain of carbon atoms
- Name this longest root chain using standard naming rules
- Name each side chain by changing the suffix of the name of the alkane from "-ane" to "-yl"
- Number the root chain so that sum of the numbers assigned to each side group will be as low as possible
- Number and name the side chains before the name of the root chain
- If there are multiple side chains of the same type, use prefixes such as "di-" and "tri-" to indicate it as such, and number each one.
- Add side chain names in alphabetical (disregarding "di-" etc. prefixes) order in front of the name of the root chain

Cyclic Alkanes

So-called cyclic alkanes are, in the technical sense, *not* alkanes, but cycloalkanes. They are hydrocarbons just like alkanes, but contain one or more rings. Simple cycloalkanes have a prefix "cyclo-" to distinguish them from alkanes. Cycloalkanes are named as per their acyclic counterparts with respect to the number of carbon atoms, e.g., cyclopentane (C_5H_{10}) is a cycloalkane with 5 carbon atoms just like pentane (C_5H_{12}), but they are joined up in a five-membered ring. In a similar manner, propane and cyclopropane, butane and cyclobutane, etc. Substituted cycloalkanes are named similar to substituted alkanes — the cycloalkane ring is stated, and the substituents are according to their position on the ring, with the numbering decided by Cahn-Ingold-Prelog rules. Cycloalkanes (also called naphthenes - not to be confused with naphthalene) are types of alkanes that have one or more rings of carbon atoms in the chemical structure of their molecules. Alkanes are types of organic hydrocarbon compounds that have only single chemical bonds in their chemical structure. Cycloalkanes consist of only carbon (C) and hydrogen (H) atoms and are saturated because there are no multiple C-C bonds to hydrogenate (add more hydrogen to). A general chemical formula for cycloalkanes would be $C_nH_{2(n+1-g)}$ where n = number of C atoms and g = number of rings in the molecule. Cycloalkanes with a single ring are named analogously to their normal alkane counterpart of the same carbon count: cyclopropane, cyclobutane, cyclopentane, cyclohexane, etc. The larger cycloalkanes, with greater than 20 carbon atoms are typically called cycloparaffins.

Cycloalkanes are classified into small, common, medium, and large cycloalkanes, where cyclopropane and cyclobutane are the small ones, cyclopentane, cyclohexane, cycloheptane are the common ones, cyclooctane through cyclotridecane are the medium ones, and the rest are the larger ones.

The naming of polycyclic alkanes such as bicyclic alkanes and spiro alkanes is more complex, with the base name indicating the number of carbons in the ring system, a prefix indicating the number of rings (e.g., "bicyclo"), and a numeric prefix before that indicating the number of carbons in each part of each ring, exclusive of vertices. For instance, a bicyclooctane that consists of a six-member ring and a four-member ring, which share two adjacent carbon atoms that form a shared edge, is [4.2.0]-bicyclooctane. That part of the six-member ring, exclusive of the shared edge has 4 carbons. That part of the four-member ring, exclusive of the shared edge, has 2 carbons. The edge itself, exclusive of the two vertices that define it, has 0 carbons. There is more than one convention (method or nomenclature) for the naming of compounds, which can be confusing for those who are just learning, and inconvenient for those who are well rehearsed in the older ways. For beginners it is best to learn IUPAC nomenclature from a source that is up to date, because this system is constantly being revised. In the above example [4.2.0]-bicyclooctane would be written bicyclo[4.2.0]octane to fit the conventions for IUPAC naming. It has then got room for an additional numerical prefix if there is the need to include details of other attachments to the molecule such as chlorine or a methyl group. Another convention for the naming of compounds is the common name, which is a shorter name and it gives less information about the compound. An example of a common name is terpineol, the name of which can tell us only that it is an alcohol (because the suffix 'ol' is in the name) and it should then have a hydroxide (OH) group attached to it. An example of the IUPAC method is given in the image to the right. In this example the base name is listed first, which indicates the total number of carbons in both rings including the carbons making up the shared edge (e.g., heptane, which means *hept* or 7 carbons, and *ane*, which indicates only single bonding between carbons). Then in front of the base name is the numerical prefix, which lists the number of carbons in each ring, excluding the carbons that are shared by each ring, plus the number of carbons on the bridge between the rings. In this case there are two rings with two carbons each and a single bridge with one carbon, excluding the carbons shared by it and the other two rings. There is a total of three numbers and they are listed in descending order separated by dots, thus: [2.2.1]. Before the numerical prefix is another prefix indicating the number of rings (e.g., "bicyclo"). Thus, the name is bicyclo[2.2.1]heptane. The group of cycloalkanes are also known as naphthenes.

Properties

Cycloalkanes are similar to alkanes in their general physical properties, but they have higher boiling points, melting points, and densities than alkanes. This is due to stronger London forces because the ring shape allows for a larger area of contact. Containing only C-C and C-H bonds, unreactivity of cycloalkanes with little or no ring strain are comparable to non-cyclic alkanes.

Ring Strain

The carbon atoms in cycloalkanes are sp^3 hybridized and are therefore a deviation from the ideal tetrahedral bond angles of 109°28'. This causes an increase in potential energy and an overall destabilizing effect. Eclipsing of hydrogen atoms is an important destabilizing effect, as well. The strain energy of a cycloalkane is the theoretical increase in energy caused by the compound's geometry, and is calculated by comparing the experimental standard enthalpy change of combustion of the cycloalkane with the value calculated using average bond energies. Ring strain is highest for cyclopropane, in which the carbon atoms form a triangle and therefore have 60 degree C-C-C bond angles. There are also three pairs of eclipsed hydrogens. The ring strain is calculated to be around 120 kJ/mol. Cyclobutane has the carbon atoms in a puckered square with approximately 90-degree bond angles; "puckering" reduces the eclipsing interactions between hydrogen atoms. Its ring strain is therefore slightly less, at around 110 kJ/mol.

For a theoretical planar cyclopentane the C-C-C bond angles would be 108 degrees, very close to the measure of the tetrahedral angle. Actual cyclopentane molecules are puckered, but this changes only the bond angles slightly so that angle strain is relatively small. The eclipsing interactions are also reduced, leaving a ring strain of about 25 kJ/mol. In cyclohexane the ring strain and eclipsing interactions are negligible because the puckering of the ring allows ideal tetrahedral bond angles to be achieved. As well, in the most stable *chair form* of cyclohexane, axial hydrogens on adjacent carbon atoms are pointed in opposite directions, virtually eliminating eclipsing strain. After cyclohexane, the molecules are unable to take a structure with no ring strain, resulting in an increase in strain energy, which peaks at 9 carbons (around 50 kJ/mol). After that, strain energy slowly decreases until 12 carbon atoms, where it drops significantly; at 14, another significant drop occurs and the strain is on a level comparable with 10 kJ/mol. After 14 carbon atoms, sources disagree on what happens to ring strain, some indicating that it increases steadily, others saying that it disappears entirely. However, bond angle strain and eclipsing strain are an issue only for smaller rings.

Reactions

The simple and the bigger cycloalkanes are very stable, like alkanes, and their reactions, for example, radical chain reactions, are like alkanes. The small cycloalkanes - in particular, cyclopropane - have a lower stability due to Baeyer strain and ring strain. They react similarly to alkenes, though they do not react in electrophilic addition, but in nucleophilic aliphatic substitution. These reactions are ring-opening reactions or ring-cleavage reactions of alkyl cycloalkanes. Cycloalkanes can be formed in a Diels-Alder reaction followed by a catalytic hydrogenation.

Trivial Names

The trivial (non-systematic) name for alkanes is "paraffins." Together, alkanes are known as the *paraffin series*. Trivial names for compounds are usually historical artifacts. They were coined before the development of systematic names, and have been retained due to familiar usage in industry. Cycloalkanes are also called naphthenes. It is almost certain that the term paraffin stems from the petrochemical industry. Branched-chain alkanes are called *isoparaffins*. The use of the term "paraffin" is a general term and often does not distinguish between pure compounds and mixtures of isomers with the same chemical formula (i.e., like a chemical anagram), e.g., pentane and isopentane.

Examples

The following trivial names are retained in the IUPAC system:

- isobutane for 2-methylpropane
- isopentane for 2-methylbutane
- neopentane for 2,2-dimethylpropane.

Melting Point

The melting points of the alkanes follow a similar trend to boiling points for the same reason as outlined above. That is, (all other things being equal) the larger the molecule the higher the melting point. There is one significant difference between boiling points and melting points. Solids have more rigid and fixed structure than liquids.

This rigid structure requires energy to break down. Thus the better put together solid structures will require more energy to break apart. For alkanes, this can be seen from the graph above (i.e., the blue line). The odd-numbered alkanes have a lower trend in melting points than even numbered alkanes. This is because even numbered alkanes pack well in the solid phase, forming a well-organised structure, which requires more energy to break apart. The odd-number alkanes pack less well and so the "looser" organised solid packing structure requires less energy to break apart.

The melting points of branched-chain alkanes can be either higher or lower than those of the corresponding straight-chain alkanes, again depending on the ability of the alkane in question to pack well in the solid phase: This is particularly true for isoalkanes (2-methyl isomers), which often have melting points higher than those of the linear analogues.

Conductivity and Solubility

Alkanes do not conduct electricity, nor are they substantially polarized by an electric field. For this reason they do not form hydrogen bonds and are insoluble in polar solvents such as water. Since the hydrogen bonds between individual water molecules are aligned away from an alkane molecule, the coexistence of an alkane and water leads to an increase in molecular order (a reduction in entropy).

As there is no significant bonding between water molecules and alkane molecules, the second law of thermodynamics suggests that this reduction in entropy should be minimised by minimising the contact between alkane and water: Alkanes are said to be hydrophobic in that they repel water.

Their solubility in nonpolar solvents is relatively good, a property that is called lipophilicity. Different alkanes are, for example, miscible in all proportions among themselves.

The density of the alkanes usually increases with increasing number of carbon atoms, but remains less than that of water. Hence, alkanes form the upper layer in an alkane-water mixture.

Special Cases

In *n*-pentane, the terminal methyl groups experience additional pentane interference.

Replacing hydrogen by fluorine in polytetrafluoroethylene changes the stereochemistry from the zigzag geometry to that of a helix due to electrostatic repulsion of the fluorine atoms in the 1,3 positions. Evidence for the helix structure in the crystalline state is derived from X-ray crystallography and from NMR spectroscopy and circular dichroism in solution.

Spectroscopic Properties

Virtually all organic compounds contain carbon – carbon and carbon – hydrogen bonds, and so show some of the features of alkanes in their spectra. Alkanes are notable for having no other groups, and therefore for the *absence* of other characteristic spectroscopic features.

Infrared Spectroscopy

The carbon–hydrogen stretching mode gives a strong absorption between 2850 and 2960 cm”[1], while the carbon–

carbon stretching mode absorbs between 800 and 1300 cm”1. The carbon–hydrogen bending modes depend on the nature of the group: methyl groups show bands at 1450 cm”1 and 1375 cm”1, while methylene groups show bands at 1465 cm”1 and 1450 cm”1. Carbon chains with more than four carbon atoms show a weak absorption at around 725 cm”1.

NMR Spectroscopy

The proton resonances of alkanes are usually found at δ_H = 0.5 – 1.5. The carbon-13 resonances depend on the number of hydrogen atoms attached to the carbon: δ_C = 8 – 30 (primary, methyl, $-CH_3$), 15 – 55 (secondary, methylene, $-CH_2-$), 20 – 60 (tertiary, methyne, C-H) and quaternary. The carbon-13 resonance of quaternary carbon atoms is characteristically weak, due to the lack of Nuclear Overhauser effect and the long relaxation time, and can be missed in weak samples, or sample that have not been run for a sufficiently long time.

Mass Spectrometry

Alkanes have a high ionization energy, and the molecular ion is usually weak. The fragmentation pattern can be difficult to interpret, but, in the case of branched chain alkanes, the carbon chain is preferentially cleaved at tertiary or quaternary carbons due to the relative stability of the resulting free radicals. The fragment resulting from the loss of a single methyl group (M”15) is often absent, and other fragment are often spaced by intervals of fourteen mass units, corresponding to sequential loss of CH_2-groups.

Chemical Properties

In general, alkanes show an alkane activity, because their C bonds are relatively stable and cannot be easily broken. Unlike most other organic compounds, they possess no functional groups.

They react only very poorly with ionic or other polar substances. The acid dissociation constant (pK_a) values of all alkanes are above 60, hence they are practically inert to acids and bases. This inertness is the source of the term *paraffins* (with the meaning here of “lacking affinity”). In crude oil the alkane molecules have remained chemically unchanged for millions of years.

However redox reactions of alkanes, in particular with oxygen and the halogens, are possible as the carbon atoms are in a strongly reduced condition; in the case of methane, the lowest possible oxidation state for carbon (–4) is reached. Reaction with oxygen (*if* an amount of the least is enough to meet the reaction stoichiometry) leads to combustion without any smoke; with halogens, substitution. In addition, alkanes have been shown to interact with, and bind to, certain transition metal complexes in.

Free radicals, molecules with unpaired electrons, play a large role in most reactions of alkanes, such as cracking and reformation where long-chain alkanes are converted into shorter-chain alkanes and straight-chain alkanes into branched-chain isomers.

In highly branched alkanes, the bond angle may differ significantly from the optimal value (109.5°) in order to allow the different groups sufficient space. This causes a tension in the molecule, known as steric hindrance, and can substantially increase the reactivity.

Cracking

Cracking breaks larger molecules into smaller ones. This can be done with a thermal or catalytic method. The thermal cracking process follows a homolytic mechanism with formation of free-radicals. The catalytic cracking process involves the presence of acid catalysts (usually solid acids such as silica-alumina and zeolites), which promote a heterolytic (asymmetric) breakage of bonds yielding pairs of ions of opposite charges, usually a carbocation and the very unstable hydride anion. Carbon-localized free-radicals and cations are both highly unstable and undergo processes of chain rearrangement, C-C scission in position beta (i.e., cracking) and intra- and intermolecular hydrogen transfer or hydride transfer. In both types of processes, the corresponding reactive intermediates (radicals, ions) are permanently regenerated, and thus they proceed by a self-propagating chain mechanism. The chain of reactions is eventually terminated by radical or ion recombination.

In petroleum geology and chemistry, cracking is the process whereby complex organic molecules such as kerogens or heavy hydrocarbons are broken down into simpler molecules such as light hydrocarbons, by the breaking of carbon-carbon bonds in the precursors. The rate of cracking and the end products are strongly dependent on the temperature and presence of catalysts. Cracking is the breakdown of a large alkane into smaller, more useful alkanes and alkenes. Simply put, hydrocarbon cracking is the process of breaking long-chain hydrocarbons into short ones. More loosely, outside the field of petroleum chemistry, the term "cracking" is used to describe any type of splitting of molecules under the influence of heat, catalysts and solvents, such as in processes of destructive distillation or pyrolysis.

The thermal cracking method (also known as "Shukhov cracking process") was invented by Russian engineer Vladimir Shukhov and patented in 1891 in the Russian empire, patent no. 12926, November 27, 1891. This process was modified by the American engineer William Merriam Burton and patented as U.S. patent 1,049,667 on June 8, 1908.

In 1924, the delegation of the American "Sinclair Oil Corporation" visited Shukhov. Sinclair Oil protested the personal right appropriated by the Rockefeller "Standard Oil" concern on the discovery of oil cracking. It indicated that Burton's patent used by the "Standard Oil" concern was the modified patent of Shukhov, proved to the Americans that the Burton's method was just the slightly changed modification of his 1891 patents. An agreement between the American companies was made not to buy the patent from the Soviet Union. Oil refinery cracking processes allow the production of "light" products such as liquefied petroleum gas (LPG) and gasoline from heavier crude oil distillation fractions such as gas oils and residues. Fluid catalytic cracking produces a high yield of gasoline and LPG, while hydrocracking is a major source of jet fuel, diesel, naphtha, and LPG. Thermal cracking is currently used to "upgrade" very heavy fractions or to produce light fractions or distillates, burner fuel and/or petroleum coke.

Chapter 10

Electrostatic Catalysis

Stabilization of charged transition states can also be by residues in the active site forming ionic bonds (or partial ionic charge interactions) with the intermediate. These bonds can either come from acidic or basic side chains found on amino acids such as lysine, arginine, aspartic acid or glutamic acid or come from metal cofactors such as zinc. Metal ions are particularly effective and can reduce the pKa of water enough to make it an effective nucleophile.

Systematic computer simulation studies established that electrostatic effects give, by far, the largest contribution to catalysis. In particular, it has been found that enzyme provides an environment which is more polar than water, and that the ionic transition states are stabilized by fixed dipoles. This is very different from transition state stabilization in water, where the water molecules must pay with "reorganisation energy". In order to stabilize ionic and charged states. Thus, the catalysis is associated with the fact that the enzyme polar groups are preorganised

Covalent Catalysis

Covalent catalysis involves the substrate forming a transient covalent bond with residues in the active site or with a cofactor. This adds an additional covalent intermediate to the reaction, and helps to reduce the energy of later transition states of the reaction.

The covalent bond must, at a later stage in the reaction, be broken to regenerate the enzyme. This mechanism is found in enzymes such as proteases like chymotrypsin and trypsin, where an acyl-enzyme intermediate is formed. Schiff base formation using the free amine from a lysine residue is another mechanism, as seen in the enzyme aldolase during glycolysis.

Some enzymes utilise non-amino acid cofactors such as pyridoxal phosphate (PLP) or thiamine pyrophosphate (TPP) to form covalent intermediates with reactant molecules. Such covalent intermediates function to reduce the energy of later transition states, similar to how covalent intermediates formed with active site amino acid residues allow stabilization, but the capabilities of cofactors allow enzymes to carryout reactions that amino acid side residues alone could not. Enzymes utilising such cofactors include the PLP-dependent enzyme aspartate transaminase and the TPP-dependent enzyme pyruvate dehydrogenase.

It is important to clarify that covalent catalysis does correspond in most cases to simply the use of a specific mechanism rather than to true catalysis. For example, the energetics of the covalent bond to the serine molecule in chymotrypsin should be compared to the well-understood covalent bond to the nucleophile in the uncatalysed solution reaction. A true proposal of a covalent catalysis (where the barrier is lower than the corresponding barrier in solution) would require, for example, a partial covalent bond to the transition state by an enzyme group (e.g., a very strong hydrogen bond), and such effects do not contribute significantly to catalysis.

Quantum Tunnelling

These traditional "over the barrier" mechanisms have been challenged in some cases by models and observations of "through the barrier" mechanisms (quantum tunnelling). Some enzymes operate with kinetics which are faster than what would be predicted by the classical $\Delta G^{\ddagger}$. In "through the barrier" models, a proton or an electron can tunnel through activation barriers. Quantum tunnelling for protons has been observed in tryptamine oxidation by aromatic amine dehydrogenase.

Interestingly, quantum tunnelling does not appear to provide a major catalytic advantage, since the tunnelling contributions are similar in the catalysed and the uncatalysed reactions in solution. However, the tunnelling contribution (typically enhancing rate constants by a factor of ~1000 compared to the rate of reaction for the classical 'over the barrier' route) is likely crucial to the viability of biological organisms. This emphasizes the general importance of tunnelling reactions in biology. In 1971-1972 the first quantum-mechanical model of enzyme catalysis was formulated.

Examples of Catalytic Mechanisms

In reality, most enzyme mechanisms involve a combination of several different types of catalysis.

Specificity

Enzymes are usually very specific as to which reactions they catalyse and the substrates that are involved in these reactions. Complementary shape, charge and hydrophilic/ hydrophobic characteristics of enzymes and substrates are responsible for this specificity. Enzymes can also show impressive levels of stereospecificity, regioselectivity and chemoselectivity. Some of the enzymes showing the highest specificity and accuracy are involved in the copying and expression of the genome. These enzymes have "proof-reading" mechanisms. Here, an enzyme such as DNA polymerase catalyses a reaction in a first step and then checks that the product is correct in a second step. This two-step process results in average error rates of less than 1 error in 100 million reactions in high-fidelity mammalian polymerases. Similar proofreading mechanisms are also found in RNA polymerase, aminoacyl tRNA synthetases and ribosomes. Some enzymes that produce secondary metabolites are described as promiscuous, as they can act on a relatively broad range of different substrates. It has been suggested that this broad substrate specificity is important for the evolution of new biosynthetic pathways.

Mechanisms

Enzymes can act in several ways, all of which lower $\Delta G^{\ddagger}$ (Gibbs energy):

- Lowering the activation energy by creating an environment in which the transition state is stabilized (e.g. straining the shape of a substrate—by binding the transition-state conformation of the substrate/product molecules, the enzyme distorts the bound substrate(s) into their transition state form, thereby reducing the amount of energy required to complete the transition).

- Lowering the energy of the transition state, but without distorting the substrate, by creating an environment with the opposite charge distribution to that of the transition state.
- Providing an alternative pathway. For example, temporarily reacting with the substrate to form an intermediate ES complex, which would be impossible in the absence of the enzyme.
- Reducing the reaction entropy change by bringing substrates together in the correct orientation to react. Considering $\Delta H^{\ddagger}$ alone overlooks this effect.
- Increases in temperatures speed up reactions. Thus, temperature increases help the enzyme function and develop the end product even faster. However, if heated too much, the enzyme's shape deteriorates and the enzyme becomes denatured. Some enzymes like thermolabile enzymes work best at low temperatures.

It is interesting that this entropic effect involves destabilization of the ground state, and its contribution to catalysis is relatively small.

Transition State Stabilization

The understanding of the origin of the reduction of $\Delta G^{\ddagger}$ requires one to find out how the enzymes can stabilize its transition state more than the transition state of the uncatalysed reaction. It seems that the most effective way for reaching large stabilization is the use of electrostatic effects, in particular, when having a relatively fixed polar environment that is oriented toward the charge distribution of the transition state. Such an environment does not exist in the uncatalysed reaction in water.

Dynamics and Function

The internal dynamics of enzymes has been suggested to be linked with their mechanism of catalysis. Internal dynamics are the movement of parts of the enzyme's structure, such as individual amino acid residues, a group of amino acids, or even an entire protein domain. These movements occur at various time-scales ranging from femtoseconds to seconds. Networks of protein residues throughout an enzyme's structure can contribute to catalysis through dynamic motions. This is simply seen in the kinetic scheme of the combined process, enzymatic activity and dynamics; this scheme can have several independent Michaelis-Menten-like reaction pathways that are connected through fluctuation rates. Protein motions are vital to many enzymes, but whether small and fast vibrations, or larger and slower conformational movements are more important depends on the type of reaction involved. However, although these movements are important in binding and releasing substrates and products, it is not clear if protein movements help to accelerate the chemical steps in enzymatic reactions. These new insights also have implications in understanding allosteric effects and developing new medicines. The presence of multiple domains in proteins gives rise to a great deal of flexibility and mobility, leading to protein domain dynamics. Domain motions can be inferred by comparing different structures of a protein (as in Database of Molecular Motions), or they can be directly observed using spectra measured by neutron spin echo spectroscopy. They can also be suggested by sampling in extensive molecular dynamics trajectories. Domain motions are important for:

- Catalysis
- Regulatory activity
- Transport of metabolites
- Formation of protein assemblies
- Cellular locomotion

One of the largest observed domain motions is the 'swivelling' mechanism in pyruvate phosphate dikinase. The phosphoinositide domain swivels between two states in order to bring a phosphate group from the active site of the nucleotide binding domain to that of the phosphoenolpyruvate/pyruvate domain. The phosphate group is moved over a distance of 45A involving a domain motion of about 100 degrees around a single residue. In enzymes, the closure of one domain onto another captures a substrate by an induced fit, allowing the reaction to take place in a controlled way.

A detailed analysis by Gerstein led to the classification of two basic types of domain motion; hinge and shear. Only a relatively small portion of the chain, namely the inter-domain linker and side chains undergo significant conformational changes upon domain rearrangement.

Hinges by Secondary Structures

A study by Hayward found that the termini of α-helices and β-sheets form hinges in a large number of cases. Many hinges were found to involve two secondary structure elements acting like hinges of a door, allowing an opening and closing motion to occur.

This can arise when two neighbouring strands within a β-sheet situated in one domain, diverge apart as they join the other domain. The two resulting termini then form the bending regions between the two domains. α-helices that preserve their hydrogen bonding network when bent are found to behave as mechanical hinges, storing 'elastic energy' that drives the closure of domains for rapid capture of a substrate.

Helical to Extended Conformation

The interconversion of helical and extended conformations at the site of a domain boundary is not uncommon. In calmodulin, torsion angles change for five residues in the middle of a domain linking α-helix. The helix is split into two, almost perpendicular, smaller helices separated by four residues of an extended strand.

Shear Motions

Shear motions involve a small sliding movement of domain interfaces, controlled by the amino acid side chains within the interface. Proteins displaying shear motions often have a layered architecture: stacking of secondary structures. The interdomain linker has merely the role of keeping the domains in close proximity.

Domain Motion and Functional Dynamics in Enzymes

The analysis of the internal dynamics of structurally different, but functionally similar enzymes has highlighted a common relationship between the positioning of the active site and the two principal protein sub-domains. In fact, for several members of the hydrolase superfamily, the catalytic site is located close to the interface separating the two principal quasi-rigid domains. Such positioning appears instrumental for maintaining the precise geometry of the active site, while allowing for an appreciable functionally oriented modulation of the flanking regions resulting from the relative motion of the two sub-domains.

Domain Definition from Structural Co-ordinates

The importance of domains as structural building blocks and elements of evolution has brought about many automated methods for their identification and classification in proteins of known structure.

Automatic procedures for reliable domain assignment is essential for the generation of the domain databases, especially as the number of protein structures is increasing. Although the boundaries of a domain can be determined by visual inspection, construction of an automated method is not straightforward.

Problems occur when faced with domains that are discontinuous or highly associated. The fact that there is no standard definition of what a domain really is has meant that domain assignments have varied enormously, with each researcher using a unique set of criteria. A structural domain is a compact, globular sub-structure with more interactions within it than with the rest of the protein. Therefore, a structural domain can be determined by two visual characteristics; its compactness and its extent of isolation. Measures of local compactness in proteins have been used in many of the early methods of domain assignment and in several of the more recent methods. One of the first algorithms used a Cα-Cα distance map together with a hierarchical clustering routine that considered proteins as several small segments, 10 residues in length. The initial segments were clustered one after another based on inter-segment distances; segments with the shortest distances were clustered and considered as single segments thereafter.

The stepwise clustering finally included the full protein. Go also exploited the fact that inter-domain distances are normally larger than intra-domain distances; all possible Cα-Cα distances were represented as diagonal plots in which there were distinct patterns for helices, extended strands and combinations of secondary structures.

The method by Sowdhamini and Blundell clusters secondary structures in a protein based on their Cα-Cα distances and identifies domains from the pattern in their dendrograms. As the procedure does not consider the protein as a continuous chain of amino acids there are no problems in treating discontinuous domains. Specific nodes in these dendrograms are identified as tertiary structural clusters of the protein, these include both super-secondary structures and domains. The DOMAK algorithm is used to create the 3Dee domain database. It calculates a 'split value' from the number of each type of contact when the protein is divided arbitrarily into two parts. This split value is large when the two parts of the structure are distinct. The method of Wodak and Janin was based on the calculated interface areas between two chain segments repeatedly cleaved at various residue positions.

Interface areas were calculated by comparing surface areas of the cleaved segments with that of the native structure. Potential domain boundaries can be identified at a site where the interface area was at a minimum. Other methods have used measures of solvent accessibility to calculate compactness.

The PUU algorithm incorporates a harmonic model used to approximate inter-domain dynamics. The underlying physical concept is that many rigid interactions will occur within each domain and loose interactions will occur between domains. This algorithm is used to define domains in the FSSP domain database.

Swindells (1995) developed a method, DETECTIVE, for identification of domains in protein structures based on the idea that domains have a hydrophobic interior. Deficiencies were found to occur when hydrophobic cores from different domains continue through the interface region.

RigidFinder is a novel method for identification of protein rigid blocks (domains and loops) from two different conformations. Rigid blocks are defined as blocks where all inter residue distances are conserved across conformations.

A general method to identify *dynamical domains*, that is protein regions that behave approximately as rigid units in the course of structural fluctuations, has been introduced by Potestio et al. and, among other applications was also used to compare the consistency of the dynamics-based domain subdivisions with standard structure-based ones. The method, termed PiSQRD, is publicly available in the form of a webserver. The latter allows users to optimally subdivide single-chain or multimeric proteins into quasi-rigid domains based on the collective modes of fluctuation of the system. By default the latter are calculated through an elastic network model; alternatively pre-calculated essential dynamical spaces can be uploaded by the user.

Example domains:

- Armadillo repeats : named after the β-catenin-like Armadillo protein of the fruit fly *Drosophila*.
- Basic Leucine zipper domain (bZIP domain) : is found in many DNA-binding eukaryotic proteins. One part of the domain contains a region that mediates sequence-specific DNA-binding properties and the Leucine zipper that is required for the dimerization of two DNA-binding regions. The DNA-binding region comprises a number of basic aminoacids such as arginine and lysine
- Cadherin repeats : Cadherins function as Ca^{2+}-dependent cell-cell adhesion proteins. Cadherin domains are extracellular regions which mediate cell-to-cell homophilic binding between cadherins on the surface of adjacent cells.
- Death effector domain (DED) : allows protein-protein binding by homotypic interactions (DED-DED). Caspase proteases trigger apoptosis via proteolytic cascades. Pro-Caspase-8 and pro-caspase-9 bind to specific adaptor molecules via DED domains and this leads to autoactivation of caspases.
- EF hand : a helix-turn-helix structural motif found in each structural domain of the signalling protein calmodulin and in the muscle protein troponin-C.
- Immunoglobulin-like domains : are found in proteins of the immunoglobulin superfamily (IgSF). They contain about 70-110 amino acids and are classified into different categories (IgV, IgC1, IgC2 and IgI) according to their size and function. They possess a characteristic fold in which two beta sheets form a "sandwich" that is

stabilized by interactions between conserved cysteines and other charged amino acids. They are important for protein-to-protein interactions in processes of cell adhesion, cell activation, and molecular recognition. These domains are commonly found in molecules with roles in the immune system.

- Phosphotyrosine-binding domain (PTB) : PTB domains usually bind to phosphorylated tyrosine residues. They are often found in signal transduction proteins. PTB-domain binding specificity is determined by residues to the amino-terminal side of the phosphotyrosine. Examples: the PTB domains of both SHC and IRS-1 bind to a NPXpY sequence. PTB-containing proteins such as SHC and IRS-1 are important for insulin responses of human cells.
- Pleckstrin homology domain (PH) : PH domains bind phosphoinositides with high affinity. Specificity for PtdIns(3)P, PtdIns(4)P, PtdIns(3,4)P2, PtdIns(4,5)P2, and PtdIns(3,4,5)P3 have all been observed. Given the fact that phosphoinositides are sequestered to various cell membranes (due to their long lipophilic tail) the PH domains usually causes recruitment of the protein in question to a membrane where the protein can exert a certain function in cell signalling, cytoskeletal reorganisation or membrane trafficking.
- Src homology 2 domain (SH2) : SH2 domains are often found in signal transduction proteins. SH2 domains confer binding to phosphorylated tyrosine (pTyr). Named after the phosphotyrosine binding domain of the src viral oncogene, which is itself a tyrosine kinase.
- Zinc finger DNA binding domain (ZnF_GATA) : ZnF_GATA domain-containing proteins are typically transcription factors that usually bind to the DNA sequence [AT]GATA[AG] of promoters.

Allosteric Modulation

Allosteric sites are sites on the enzyme that bind to molecules in the cellular environment. The sites form weak, noncovalent bonds with these molecules, causing a change in the conformation of the enzyme. This change in conformation translates to the active site, which then affects the reaction rate of the enzyme. Allosteric interactions can both inhibit and activate enzymes and are a common way that enzymes are controlled in the body.

Cofactors and Coenzymes

Cofactors

Some enzymes do not need any additional components to show full activity. However, others require non-protein molecules called cofactors to be bound for activity. Cofactors can be either inorganic (e.g., metal ions and iron-sulfur clusters) or organic compounds (e.g., flavin and heme).

Organic cofactors can be either prosthetic groups, which are tightly bound to an enzyme, or coenzymes, which are released from the enzyme's active site during the reaction. Coenzymes include NADH, NADPH and adenosine triphosphate. These molecules transfer chemical groups between enzymes.

An example of an enzyme that contains a cofactor is carbonic anhydrase, and is shown in the ribbon diagram above with a zinc cofactor bound as part of its active site. These tightly bound molecules are usually found in the active site and are involved in catalysis. For example, flavin and heme cofactors are often involved in redox reactions.

Enzymes that require a cofactor but do not have one bound are called *apoenzymes* or *apoproteins*. An apoenzyme together with its cofactor(s) is called a *holoenzyme* (this is the active form). Most cofactors are not covalently attached to an enzyme, but are very tightly bound. However, organic prosthetic groups can be covalently bound (e.g., biotin in the enzyme pyruvate carboxylase). The term "holoenzyme" can also be applied to enzymes that contain multiple protein subunits, such as the DNA polymerases; here the holoenzyme is the complete complex containing all the subunits needed for activity.

A cofactor is a non-protein chemical compound that is bound to a protein and is required for the protein's biological activity. These proteins are commonly enzymes, and cofactors can be considered "helper molecules" that assist in biochemical transformations. Cofactors are either organic or inorganic. They can also be classified depending on how tightly they bind to an enzyme, with loosely-bound cofactors termed coenzymes and tightly-bound cofactors termed prosthetic groups. Some sources also limit the use of the term "cofactor" to inorganic substances. An inactive enzyme, without the cofactor is called an apoenzyme, while the complete enzyme with cofactor is the holoenzyme.

Some enzymes or enzyme complexes require several cofactors. For example, the multienzyme complex pyruvate dehydrogenase at the junction of glycolysis and the citric acid cycle requires five organic cofactors and one metal ion: loosely bound thiamine pyrophosphate (TPP), covalently bound lipoamide and flavin adenine dinucleotide (FAD), and the cosubstrates nicotinamide adenine dinucleotide (NAD^+) and coenzyme A (CoA), and a metal ion (Mg^{2+}). Organic cofactors are often vitamins or are made from vitamins. Many contain the nucleotide adenosine monophosphate (AMP) as part of their structures, such as ATP, coenzyme A, FAD, and NAD^+. This common structure may reflect a common evolutionary origin as part of ribozymes in an ancient RNA world. It has been suggested that the AMP part of the molecule can be considered a kind of "handle" by which the enzyme can "grasp" the coenzyme to switch it between different catalytic centres. Cofactors can be divided into two broad groups: organic cofactors, such as flavin or heme, and inorganic cofactors, such as the metal ions Mg^{2+}, Cu^+, Mn^{2+}, or iron-sulfur clusters.

Organic cofactors are sometimes further divided into *coenzymes* and *prosthetic groups*. The term coenzyme refers specifically to enzymes and, as such, to the functional properties of a protein. On the other hand, "prosthetic group" emphasizes the nature of the binding of a cofactor to a protein (tight or covalent) and, thus, refers to a structural property. Different sources give slightly different definitions of coenzymes, cofactors, and prosthetic groups. Some consider tightly-bound organic molecules as prosthetic groups and not as coenzymes, while others define all non-protein organic molecules needed for enzyme activity as coenzymes, and classify those that are tightly bound as coenzyme prosthetic groups. It should be noted that these terms are often used loosely. A 1979 letter in *Trends in Biochemical Sciences* noted the confusion in the literature and the essentially arbitrary distinction made between prosthetic groups and coenzymes and proposed the following scheme.

Here, cofactors were defined as an additional substance apart from protein and substrate that is required for enzyme activity and a prosthetic group as a substance that undergoes its whole catalytic cycle attached to a single enzyme molecule. However, the author could not arrive at a single all-encompassing definition of a "coenzyme" and proposed that this term be dropped from use in the literature. Co-factor is a component other than protein portion of many enzymes.

Inorganic

Metal Ions

Metal ions are common cofactors. The study of these cofactors falls under the area of bioinorganic chemistry. In nutrition, the list of essential trace elements reflects their role as cofactors. In humans this list commonly includes iron, magnesium, manganese, cobalt, copper, zinc, selenium, and molybdenum.

Although chromium deficiency causes impaired glucose tolerance, no human enzyme that uses this metal as a cofactor has been identified. Iodine is also an essential trace element, but this element is used as part of the structure of thyroid hormones rather than as an enzyme cofactor.

Calcium is another special case, in that it is required as a component of the human diet, and it is needed for the full activity of many enzymes, such as nitric oxide synthase, protein phosphatases, and adenylate kinase, but calcium activates these enzymes in allosteric regulation, often binding to these enzymes in a complex with calmodulin. Calcium is, therefore, a cell signalling molecule, and not usually considered a cofactor of the enzymes it regulates.

Other organisms require additional metals as enzyme cofactors, such as vanadium in the nitrogenase of the nitrogen-fixing bacteria of the genus *Azotobacter*, tungsten in the aldehyde ferredoxin oxidoreductase of the thermophilic archaean *Pyrococcus furiosus*, and even cadmium in the carbonic anhydrase from the marine diatom *Thalassiosira weissflogii*.

In many cases, the cofactor includes both an inorganic and organic component. One diverse set of examples is the haem proteins, which consist of a porphyrin ring coordinated to iron.

Metalloenzymes are Widespread and Diverse in Function

It is estimated that approximately half of all proteins contain a metal. In another estimate, about one quarter to one third of all proteins are proposed require metals to carry out their functions. Thus, metalloproteins have many different functions in cells, such as enzymes, transport and storage proteins, and signal transduction proteins.

Coordination Chemistry Principles

In metalloproteins, metal ions are usually coordinated by nitrogen, oxygen or sulfur centres belonging to amino acid residues of the protein. These donor groups are often provided by side-chains on the amino acid residues. Especially important are the imidazole substituent in histidine residues, thiolate substituents in cysteinyl residues, and carboxylate groups provided by aspartate. Given the diversity of the metalloproteome, virtually all amino acid residues have been shown to bind metal centres. The peptide backbone also provides donor groups, these include deprotonated amides and the amide carbonyl oxygen centres. In addition to donor groups that are provided by amino acid residues, a large number of organic cofactors function as ligands. Perhaps most famous are the tetradentate N4 macrocyclic ligands incorporated into the heme protein. Inorganic ligands such as sulfide and oxide are also common.

Storage and Transport Metalloproteins

Oxygen Carriers

Hemoglobin, which is the principal oxygen carrier in humans has four sub-units in which the iron(II) ion is coordinated by the planar, macrocyclic ligand protoporphyrin IX (PIX) and the imidazole nitrogen atom of a histidine residue. The sixth coordination site contains a water molecule or a dioxygen molecule. Myoglobin has only one such unit. The active site is located in an hydrophobic pocket. This is important as, without it, the iron(II) would be irreversibly oxidised to iron(III). The equilibrium constant for the formation of HbO_2 is such that oxygen is taken up or released depending on the partial pressure of oxygen in the lungs or in muscle. In hemoglobin the four sub-units show a cooperativity effect which allows for easy oxygen transfer from hemoglobin to myoglobin. In both hemoglobin and myoglobin it is sometimes incorrectly stated that the oxygenated species contains iron(III). It is now known that the diamagnetic nature of these species is due to the fact that the iron(II) is in the low-spin state. In oxyhemoglobin the iron atom is located in the plane of the porphyrin ring, but in the paramagnetic deoxyhemoglobin the iron atom lies above the plane of the ring. The change in spin state is a cooperative effect of higher crystal field splitting and smaller ionic radius of Fe^{2+} in the oxy- moiety.

Hemerythrin is another iron-containing oxygen carrier. The oxygen binding site is a binuclear iron centre. The iron atoms are coordinated to the protein through the carboxylate side chains of a glutamate and aspartate and five histidine residues. The uptake of O_2 by hemerythrin is accompanied by two-electron oxidation of the reduced binuclear centre to produce bound peroxide (OOH-). The mechanism of oxygen uptake and release have been worked out in detail. Hemocyanins carry oxygen in the blood of most molluscs, and some arthropods such as the horseshoe crab. They are second only to hemoglobin in biological popularity of use in oxygen transport. On oxygenation the two copper(I) atoms at the active site are oxidised to copper(II) and the dioxygen molecules is reduced to peroxide, O_2^{2-}.

Bibliography

Arthur, Paul: *Lecture Demonstrations in General Chemistry,* McGraw-Hill Book Company Inc, New York, 1939.

Barber, Jacqueline: *Chemical Reactions,* GEMS Publishing, Berkeley, CA, 1999.

Becker, Robert: *Twenty Demonstrations Guaranteed to Knock Your Socks Off!* Flinn Scientific, Inc., 1994.

Chandler, David: *Introduction to Modern Statistical Mechanics*, Oxford University Press, New York, 1987.

Christova, R., S. Alexandrov, D. Tsalev, B. Jeljazkova, V. Mihajlova: *Practice in Quantitative Analysis*, Ohridski University Press, Sofia, 2003.

Denbigh, K.: *The Principles of Chemical Equilibrium*, Cambridge Univ. Press, London, 1966.

Dodd, J.S.: *The ACS Style Guide*, 2nd. ed.; ACS: Washington, D.C., 1997,

Engelmann, S., & Carnine, D.: *Theory of Instruction: Principles and Applications*, ADI Press, Eugene, OR, 1991.

Evans, A.: *Potentiometry and Ion Selective Electrodes*, Wiley, New York, 1987.

Flory, Paul J.: *Principles of Polymer Chemistry*, Cornell University Press, UK, 1953.

George B. Hogaboom: *Principles of Electroplating and Electroforming (Electrotyping)*, McGraw-Hill, New York, 1949.

Gordon, Lynn.: *52 Amazing Science Experiments,* Chronicle Books, San Francisco, California, 1998.

Hardwoody, L. M. and C. J. Moody: *Experimental Organic Chemistry. Standard and Microscale*, Blackwell Scientific Publications, Oxford, 1999.

Havezov, I.P.; D.L. Tsalev: *Atomic Absorption Analysis*, Nauka i Izkustvo, Sofia, 1980.

Jones, Denny: *Principles and Prevention of Corrosion,* Upper Saddle River, Prentice Hall, New Jersey, 1996.

Jones, Harry A.: *Machine Shop Practice,* Thomas Nelson and Sons, Toronto, 1941.

Kenneth I. Ozomwna: *Recent Advances in Analytical Electrochemistry 2007*, Transworld Research Network, 2007.

Lindon, John C.: *Encyclopedia of Spectroscopy and Specrometry*, Academic Press, Uk, 1999.

Linus Pauling: *The Nature of the Chemical Bond and the Structure of Molecules and Crystals; An Introduction to Modern Structural Chemistry*, Cornell University Press; Ithaca, N.Y., London, 1940.

Miller, J. N. and J. C. Miller: *Statistics and Chemometrics for Analytical Chemistry*, Prentice Hall, UK, 2000.

Nikolov, G.; P.R. Bontchev: *Mechanisms of Inorganic Reactions in Solutions*, Nauka i Izkustvo, Sofia, 1969.

Olagoke Olabisi: *Handbook of Thermoplastics,* Marcel Dekker, New York, 1997.

Porter, R.N.: S*olid State Chemistry: An Introduction*, Chapman and Hall

Prichard, E.: *Analytical Measurement Terminology*, LGC/Royal Society of Chemistry, Cambridge, UK, 2001.

Reid, R. C., J. M. Prausnitz, and B. E. Poling: *The Properties of Gases and Liquids*, McGraw-Hill, New York, 1987.

Schmidt, Lanny D.: *The Engineering of Chemical Reactions*, Oxford University Press, New York, 2005.

Seeman, N.C.: *Biomolecular Stereodynamics*, Adenine Press, New York, 1981.

Trevor H. : *Transforming Matter: A History of Chemistry from Alchemy to the Buckyball,* Baltimore: Johns Hopkins University Press, 2000.

Tsalev, D.L.; I.P. Havezov: *Atomic Absorption Analysis*, Khimia, Leningrad, 1983.

Wang, Joseph C.: *Analytical Electrochemistry*, John Wiley & Sons, Chichester, 2000.

Woodburn, J.H, Sanford, L.R., Brawley Jr., W.J.: *Demonstrations and Activities For High School Chemistry,* Parker Publishing Company Inc., West Nyack, NY, 1971.

Index

□□□